IEE CONTROL ENGINEERING SERIES 1

SERIES EDITORS: G.A. MONTGOMERIE
PROF. H. NICHOLSON

Multivariable control theory

IEE CONTROL ENGINEERING SERIES 1

Multivariable control theory

J.M. LAYTON, B.A., B.Sc. (Eng.)
Formerly Lecturer,
Department of Electronic & Electrical Engineering,
University of Birmingham,
Edgbaston,
Birmingham, England

PETER PEREGRINUS LTD.
on behalf of the
Institution of Electrical Engineers

Published by Peter Peregrinus Ltd.,
Southgate House, Stevenage, Herts. SG1 1HQ, England

First published 1976

ISBN: 0 901223 89 1

Typeset by EWC Wilkins Ltd., London and Northampton
Printed in England by A. Wheaton & Co. Exeter

Contents

Introduction

It is always dangerous to write a textbook dealing with a branch of technology that is still developing rapidly. Nevertheless, I feel that the tempo of development in the theory of control engineering has sufficiently diminished in the last few years to justify a textbook that will bring the reader up to date on the essentials of recent developments and thus avoid the necessity of consulting a large number of disseminated papers on the various topics covered.

This book is an introductory text primarily aimed at the M.Sc. student in any engineering discipline, though it should not be out of the reach of many final-year undergraduate students of control engineering. Since it deals with multivariable control theory, it nevertheless assumes that the reader has had some grounding in the elementary principles of automatic control as applied to one-input/one-output control systems. As far as mathematical requirements are concerned, the book assumes a fairly thorough knowledge of matrix theory (covering eigenvalues, eigenvectors, rank, zeros and poles of s-matrices etc.) and an elementary knowledge of the thoery of functions of a complex variable (poles and zeros, the concept of mapping etc.)

This book deals with deterministic systems and does not include any analysis of stochastic phenomena although, naturally, from time to time, the general ideas of minimising the influence of 'noise' or other disturbances is discussed.

As stated in the title, this book is about the theory, rather than the practice, of control systems. There is, in this connection, a rather apposite dictum, attributed to Lenin: 'Theory without practice is sterile, practice without theory is blind'. I agree with this dictum and indeed regret the dichotomy which, for many years, has existed between the relatively advanced level of theory in the control field and the relatively slow tempo of the implementation of this theory in engineering practice. Yet Lenin wrote a number of books of a purely theoretical nature! It is to be hoped that this book, and similar theoretical books, may lead budding practicing engineers to a better understanding of the systems that they are trying to improve and of the methods which they should use to bring about this improvement.

The book is divided into three parts. The first deals with those properties of systems that are relevant to control system theory, where the systems concerned are not necessarily control systems. The second part is a summary of a number of ways suggested by various authors to design satisfactory controllers, i.e. controllers that adequately satisfy the various performance criteria imposed by the designer. The third part deals with various aspects of optimisation theory.

It may be noted that the order of the last two parts is a reversal of the historical order of development; for during and immediately after the last war, largely due to the development of guided missiles and subsequently of space travel, the theory of optimisation gained a marked ascendancy; it was only later that the gradually increasing complexity of industrial plants required the development of humbler, 'bread and butter' multivariable control systems. I feel it is more logical, however, to precede the search for the best by the search for the good.

In conclusion, I should like to express my gratitude to my colleague, Professor H.A. Prime, who, in spite of his multifarious duties, has always found time to peruse and comment upon my typescripts.

J.M. Layton
Birmingham

A note on symbols

Any matrix, other than a row- or column-matrix, is normally represented by a bold-type, upper-casing letter: e.g. $\boldsymbol{A}$.

A column-matrix is represented by a bold-type lower-casing letter: e.g. $\boldsymbol{x}$.

The transpose of a matrix is indicated by a prime; e.g. $\boldsymbol{A}'$ is the transpose of $\boldsymbol{A}$; $\boldsymbol{x}'$ is the transpose of $\boldsymbol{x}$ and is therefore a row-matrix.

Inversion is normally represented by the index (−1): e.g. $\boldsymbol{A}^{-1}$ is the inverse of $\boldsymbol{A}$. In Chapter 8, however, we use the Rosenbrock notation $\hat{\boldsymbol{A}}$ for the inverse of $\boldsymbol{A}$.

The symbol $\boldsymbol{u}$ or $\boldsymbol{u}(t)$ has been used to denote the input-vector to a plant, expressed as a column-matrix. In view of this we have preferred to denote the output-vector by $\boldsymbol{v}$ or $\boldsymbol{v}(t)$, rather than by $\boldsymbol{y}$ or $\boldsymbol{y}(t)$, and to use the letters $\boldsymbol{x}, \boldsymbol{y}$ and $\boldsymbol{z}$ to denote various forms of the state-vector.

Part 1

Some properties of systems

Chapter 1

Systems and system representation

1.1 Introduction

The use of the word 'system' implies two essential properties: the first is the concept of interaction between a set of entities contained within the system considered, the second–almost implicit in the first–is the concept of a boundary, real or imagined, separating entities inside the system from entities outside.

In dealing with physical systems, the entities of interest are the magnitudes, usually varying with time, of certain physical quantities present within the system, these magnitudes being inter-related by physical laws. We may, moreover, be interested in the effect upon these magnitudes of agencies outside the system, external inputs to the system considered. These external inputs, while affecting the behaviour of the system quantities, may themselves not be reciprocally affected by the system quantities, and may therefore be arbitrary in their time behaviour.

In dealing with control systems in particular, our main interest will be focused upon only some of the system quantities, namely those whose behaviour we wish to control (i.e. to compel to behave with time in some predetermined manner). These quantities, which are normally accessible for purposes of measurement, we shall call output quantities. (Note that *out*puts are *in*side the system, *in*puts are *out*side!)

Finally, it is important to note that the boundaries of the system are completely elastic: we may choose, at any stage in the analysis, to consider only part of the original system as a system in its own rights or, conversely, we may choose to expand the boundaries of the original system to include new quantities, new pieces of apparatus etc. It is important, however, to remember at every stage which system is under consideration.

1.1.1 Mathematical modelling

Whatever the system considered, no quantitative analysis is possible until the various inter-relations existing between the system quantities themselves, and between these quantities and certain specified external inputs, have been expressed in mathematical terms. This very important preliminary process is called mathematical modelling. Almost invariably it leads to a compromise between accuracy of representation and tractability of mathematical form. As a first attempt, the tendency will be to make possibly severe approximations to physical reality in order to secure a simple mathematical form; after obtaining some idea of system behaviour on the basis of this first crude model, it may be felt desirable to work with a more accurate model. But at all stages it is vital to remember the nature of the approximations made in any particular model.

One early approximation which is usually made, sometimes almost unconsciously, is to consider a system with distributed parameters as an equivalent system with lumped parameters. (If, for instance, it is required to control the temperature of a room, it will first be assumed that the room is an isothermal; to do otherwise would lead to very complicated partial differential equations representing the heat flow, both conductive and convective, between any two points in the room). If this is done, the inter-relations present in almost all physical systems may be put in the form of a set of differential equations, linear or non-linear, time-dependent or time-independent (autonomous).

The order of these differential equations will depend upon the extent to which we break down the various causatory chains present in the system; the more detailed this breakdown, the larger the number of quantities involved, but the lower the order of the differential equations relating successive quantities. Theoretically, at any rate, we may visualise at one end of the scale the elimination from the equations of all system quantities except the outputs, and obtain a relatively small number of relatively high-order differential equations relating these outputs to the external inputs. In general, even if the elimination is feasible, this is not a constructive process, but if the system representation is both linear and autonomous, then, through Laplace transformation, the process leads to the very important field of control theory based upon transfer functions.

We prefer, at first, to go to the other end of the scale, and to assume that the causatory chains have been broken down sufficiently for the order of every differential equation relating any quantity to its neighbours in the causatory chain to be of order unity; alternatively, if the physical complexity of the system prevents this complete breakdown but if, possibly experimentally, we can ascertain a higher order differential equation relating any two quantities, then the breakdown may be completed as a purely mathematical exercise by expressing this higher order differential equation as an equivalent number of first-order equations–at the cost, of course, of inserting more dependent variables. (It is established that this mathematical exercise can always be carried out in an infinite number of ways [see Appendix to this Chapter] when the equation is linear and autonomous; it is normally assumed that it is feasible if it is neither.)

The breakdown of the system relations to first-order differential equations enables a systematic process of solution to be devised and, even when the equations are nonlinear, facilitates the computerisation of the solutions.

With these preliminaries, we are in a position to begin a process of analysis.

1.1.2 Canonical system equations

We consider the system as being excited by m external inputs, $u_1(t)$, $u_2(t), \ldots, u_m(t)$, applied at various points of the system. Within the system we visualise a number of quantities $x_1(t), x_2(t), \ldots, x_n(t)$, sufficient in number to allow the inter-relations of the system to be expressed as first-order differential equations. (Note that since the system has therefore been broken down as far as possible, the number of input points cannot exceed the number of system quantities, hence $m \leqslant n$.) Some of the system quantities contribute to, or may even be identical with, the outputs of the system, denoted by $v_1(t), v_2(t), \ldots, v_p(t)$, where $p \leqslant n$. Then, on the basis of the arguments of the previous section, we assume that the inter-relations between the x's and the u's can be written in the form

$$\frac{dx_r}{dt} \equiv \dot{x}_r = f_r(x_1, x_2, \ldots, x_n; u_1, u_2, \ldots, u_m; t) \qquad \text{with} \quad r = 1, 2, \ldots, n \tag{1.1}$$

where the f_r denote possibly nonlinear functions of their several arguments. The time t will be absent from these arguments in autonomous systems. These equations are written more concisely as a single column-matrix equation:

$$x = f(x, u, t) \tag{1.2}$$

where

$x = [x_1, x_2, \ldots, x_n]'$ is called the *state (column) vector,*

$u = [u_1, u_2, \ldots, u_m]'$ is called the *input (column) vector,*

this equation being the canonical (or standard) form of the *state equation* for a nonlinear, nonautonomous system.

If we suppose that the outputs are merely a subset of the state quantities, then the *output equation* assumes the simple form

$$v = Cx$$

where $v = [v_1, v_2, \ldots, v_p]'$ is the *output (column) vector,*

and C is a matrix of order $p \times n$, such that each row contains one and only one unity element, all others being zero. Note, however, that although the state equation is a first-order differential equation for x, no derivative of u is present: in some systems the elimination of this derivative may lead to a new choice of state-elements, which may not only complicate the form of C but introduce functions of the inputs into the output equation. This is illustrated later in connection with linear systems.

The process of *solving* the state equation is understood to mean finding $x(t)$ at any time $t = t_1$, given both the value of $x(t)$ at some earlier time t_0 and the value of $u(t)$ at all times in the range $t_0 \leqslant t \leqslant t_1$. The overwhelming majority of nonlinear differential equations are not soluble analytically; if, therefore, the state equations are nonlinear (more precisely if any of the f_r are nonlinear functions of the elements of x) then the best way to solve them is by computer, using step-by-step methods. If the f_r are all linear functions of the state elements, but if any of the coefficients of these state elements are functions of the inputs or of time, then the state equation takes the form of linear differential equations with time-varying coefficients, and their solution may still be an arduous process. If, however, the f_r are linear in the state elements and the coefficients of the state elements are constants, so that every f_r is of the form $f_r = \sum_{s=1}^{n} k_{rs}x_s + g_r(u, t)$, where all k_{rs} are constant, then the state equations may be solved analytically in several ways. These considerations lead to the possibility of linearising the state equation.

1.2 Linearisation of the state equation

A large majority of physical systems are linear, or very nearly so, over a wide range of amplitudes of the physical quantities involved. Even for those systems containing non-negligible nonlinearities in their normal range of operation, a linearised form of their state equation will be a valid approximation, provided that the quantities involved do not vary too widely from certain datum values about which linearisation takes place. Moreover, the results obtained from such a linear model, coupled with a knowledge of the type of nonlinearity present, should give at any rate a qualitative idea of the way in which system behaviour will differ from model behaviour. For these reasons much of control literature—and most of this book—is concerned with the theory of linear systems or, more correctly, of linear models of systems.

Consider a set of values $\boldsymbol{x}(t) = \boldsymbol{x}_d(t)$, $\boldsymbol{u}(t) = \boldsymbol{u}_d(t)$, satisfying eqn. 1.2; then

$$\dot{\boldsymbol{x}}_d = \boldsymbol{f}(\boldsymbol{x}_d, \boldsymbol{u}_d, t) \tag{1.3}$$

Consider another set of values $\boldsymbol{x}(t) = \boldsymbol{x}_d(t) + \boldsymbol{x}_i(t)$, $\boldsymbol{u}(t) = \boldsymbol{u}_d(t) + \boldsymbol{u}_i(t)$, also satisfying eqn. 1.2 and differing from the previous set by small incremental values $\boldsymbol{x}_i(t)$, $\boldsymbol{u}_i(t)$. Then

$$\dot{\boldsymbol{x}}_d + \dot{\boldsymbol{x}}_i = \boldsymbol{f}(\boldsymbol{x}_d + \boldsymbol{x}_i, \boldsymbol{u}_d + \boldsymbol{u}_i, t) \tag{1.4}$$

Assuming that $\boldsymbol{f}$ is differentiable with respect to its arguments, the right hand side of this equation may be expanded as a Taylor series, and if $\boldsymbol{x}_i$, $\boldsymbol{u}_i$, are sufficiently small in norm, we may neglect squares and higher powers of their elements. Hence, considering for simplicity the rth scalar equation in eqn. 1.4

$$\dot{x}_{dr} + \dot{x}_{ir} \doteqdot f_r(\boldsymbol{x}_d, \boldsymbol{u}_d, t) + \sum_{s=1}^{n} \frac{\partial f_r}{\partial x_s} x_{is} + \sum_{s=1}^{m} \frac{\partial f_r}{\partial u_s} u_{is},$$

$$r = 1, 2, \ldots, n$$

or, on subtracting the rth member of eqn. 1.3,

$$\dot{x}_{ir} \doteqdot \sum_{s=1}^{n} \frac{\partial f_r}{\partial x_s} x_{is} + \sum_{s=1}^{m} \frac{\partial f_r}{\partial u_s} u_{is}, \qquad r = 1, 2, \ldots, n$$

These equations may conveniently be written as a single column-matrix equation

$$\dot{\boldsymbol{x}}_i = \boldsymbol{A}\boldsymbol{x}_i + \boldsymbol{B}\boldsymbol{u}_i$$

where

$$\boldsymbol{A} = [a_{rs}] = \left[\frac{\partial f_r}{\partial x_s}\right] \text{ and is square of order } n$$

and

$$\boldsymbol{B} = [b_{rs}] = \left[\frac{\partial f_r}{\partial u_s}\right] \text{ and is of order } n \times m,$$

all partial derivatives being evaluated, by the rules of Taylor expansion, at the datum values $\boldsymbol{x} = \boldsymbol{x}_d, \boldsymbol{u} = \boldsymbol{u}_d$. Thus, treating $\boldsymbol{x}_i$ and $\boldsymbol{u}_i$ as new state and input vectors, and dropping the suffix, the linearised state equation is

$$\dot{\boldsymbol{x}} = \boldsymbol{A}\boldsymbol{x} + \boldsymbol{B}\boldsymbol{u} \tag{1.5}$$

The matrix with typical element $\dfrac{\partial f_r}{\partial x_s}$ (*before* putting $\boldsymbol{x} = \boldsymbol{x}_d, \boldsymbol{u} = \boldsymbol{u}_d$) is the derivative of the vector $\boldsymbol{f}$ with respect to the vector $\boldsymbol{x}$, in the sense that

$$d\boldsymbol{f} = \left[\frac{\partial f_r}{\partial x_s}\right] \cdot d\boldsymbol{x}$$

It is known as the Jacobian matrix relating $\boldsymbol{f}$ to $\boldsymbol{x}$, and will here be denoted by

$$J(\boldsymbol{f}:\boldsymbol{x}) \equiv \left[\frac{\partial f_r}{\partial x_s}\right] \text{ so that } A = J(\boldsymbol{f}:\boldsymbol{x}) \text{ evaluated at } \boldsymbol{x} = \boldsymbol{x}_d, \boldsymbol{u} = \boldsymbol{u}_d.$$

Similarly, (1.6)

$$J(\boldsymbol{f}:\boldsymbol{u}) \equiv \left[\frac{\partial f_r}{\partial u_s}\right] \text{ and } \quad \boldsymbol{B} = J(\boldsymbol{f}:\boldsymbol{u}) \text{ evaluated at } \boldsymbol{x} = \boldsymbol{x}_d, \boldsymbol{u} = \boldsymbol{u}_d.$$

1.2.1 Comments

(i) In a nonautonomous system, where $\boldsymbol{f}$ is explicitly a function of t, it is highly probable that $\dfrac{\partial \boldsymbol{f}}{\partial \boldsymbol{x}}$ and $\dfrac{\partial \boldsymbol{f}}{\partial \boldsymbol{u}}$ are also functions of t, and that A and B are similarly explicit functions of t. This makes the *analytical* solution of the state equation impossible except in some very special cases. Nonautonomous systems are, however, very rare.

(ii) It has been shown that, provided the incremental values $\boldsymbol{x}_i, \boldsymbol{u}_i$, are sufficiently small to justify the truncation of the Taylor series, then these incremental values satisfy the linear equation, eqn. 1.5. It has not been assumed that the datum vectors $\boldsymbol{x}_d$ and $\boldsymbol{u}_d$ are constant. It would clearly be simpler, in assessing the ranges of $\boldsymbol{x}$ and $\boldsymbol{u}$ required to validate the linear equation, to refer these ranges to constant rather than time-varying datum values. But this is not always possible, even in autonomous systems.

(iii) The presence of t as an explicit variable in the state equation is often discussed under the heading of 'time-varying parameters', a parameter in this context being some coefficient in the state differential equations. In many cases, however, the variation of the parameter with time is not *ex*plicit but *im*plicit, in the sense that the parameter in question is a function of either state or input elements, which of course in general vary with time. For instance, in the case of a jet-propelled craft, the mass varies with time because fuel is being consumed and the rate of fuel consumption may be taken as the external input; but clearly the mass does not vary *explicitly* with time, otherwise, if the fuel were shut off the mass would still vary. In this and analogous problems, the variable parameter may be expressed in terms of $\boldsymbol{x}$ and/or $\boldsymbol{u}$; alternatively, the state equations may sometimes be made more elegant mathematically by making the parameter an extra element of the state vector. In either case the state equations for the system will be autonomous. This situation will be illustrated by a simple example which will also serve to show that, even with autonomous state equations, it may be impossible to find useful, constant datum values for the purpose of linearisation.

Example A jet-propelled missile moves in the earth's gravitational field. To avoid the –in this context, irrelevant–complications of 3-dimensional motion, assume that the missile follows a radially-outward path through the earth's centre of attraction, the jet thrust being also along this radius. The equation of motion may be written:

$$m\ddot{r} = T(q,r) - h(r,\dot{r}) - Km/r^2$$

where

- $r(t)$ is the distance from the gravitational centre
- $q(t)$ is the mass rate of fuel consumption ($\geqslant 0$)
- $T(q,r)$ is the thrust, mainly dictated by q and such that $T(0,r) = 0$, but possibly modified by the atmosphere and hence by r
- $h(r,\dot{r})$ is the atmospheric drag force, such that $h(r,0) = 0$
- $m(t)$ is the missile mass (> 0)
- K is a gravitational constant.

Write this equation in state form with $r = x_1, \dot{r} = x_2, q = u$:

$$\dot{x}_1 = x_2, \; \dot{x}_2 = T/m - h(x_1,x_2)/m - K/x_1^2$$

Moreover $\dot{m} = -u$ or, on integration, $m = m_0 - \int_{t_0}^{t} u(t)dt$, where m_0 is the initial mass of the missile at launching, say $t = t_0$.

If we choose, we may substitute this value of m in the equation for $\dot{x}_2$, resulting in a very intractable function of u. If constant datum values of x_1, x_2 and u are to exist, we must have $x_2 = 0$ (and hence

$h = 0$). Also, $\dot{x}_2$ must vanish, so that the thrust must just balance the gravitational force, producing zero acceleration: $T(u, x_1) = Km/x_1^2$. Since u and x_1 are to be constant, T must be constant and hence m must be constant; hence $u = \dot{m}$ must vanish, hence T must vanish, hence $x_1 \to \infty$. Thus the only constant datum values permissible correspond to an infinite distance from the earth, zero velocity and zero thrust (zero fuel consumption). Since the deviations from the datum values must be small, it is obvious that the resulting linear equations will not serve any useful purpose in analysing the motion of the missile!

We may, alternatively, write $m = x_3$ so that we have three state equations: $\dot{x}_1 = x_2, \dot{x}_2 = T(u, x_1)/x_3 - h(x_1, x_2)/x_3 - K/x_1^2$ and $\dot{x}_3 = -u$. Whereas these equations are somewhat more elegant in form, since they avoid the integral expression for the mass, it will be found that datum values, if constant, are exactly the same as before. Clearly the system equations do not lend themselves to linearisation and are best solved as nonlinear equations using computational methods.

1.2.2 Simulation

The representation of the system by the state equation, equivalent to n scalar *first-order* differential equations, lends itself readily to physical interpretation or simulation; note that this is the reverse process to mathematical modelling: we are now trying, for experimental purposes to give a physical form to the mathematical model of the original system.

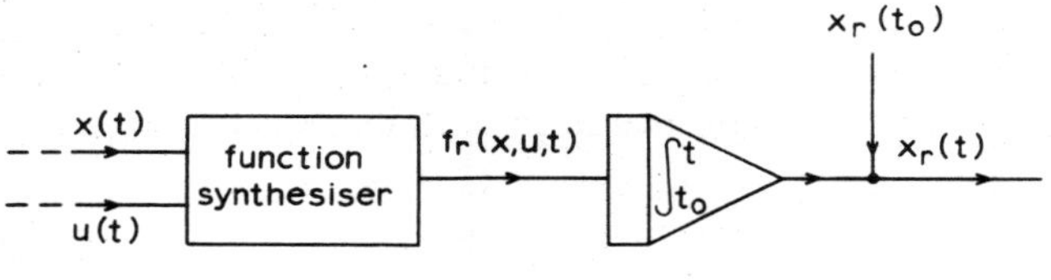

Fig. 1.1 Part schematic of simulator (general case)

Indeed if, in the general case corresponding to eqn. 1.2, $\dot{x}_r = f_r(\boldsymbol{x}, \boldsymbol{u}, t)$, then $x_r(t)$ may be simulated (physically realised) by the simple integrator circuit of Fig. 1.1; for, by integration

$$x_r(t) = x_r(t_0) + \int_{t_0}^{t} f_r\{\boldsymbol{x}(\tau), \boldsymbol{u}(\tau), \tau\} d\tau \qquad (r = 1, 2, \ldots, n)$$

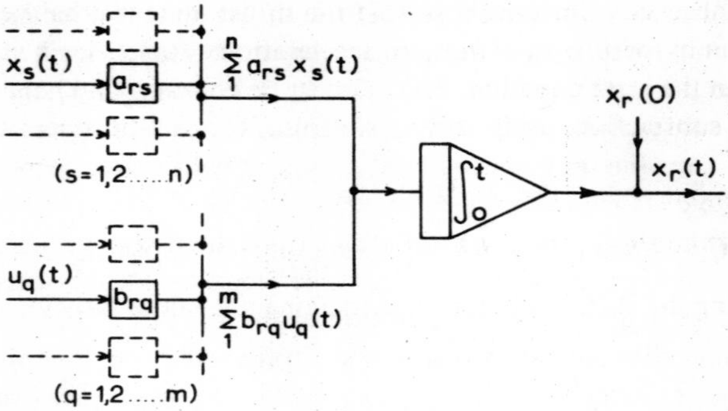

Fig. 1.2 Part schematic of simulator (autonomous, linear system)

The input f_r to any of these integrators has to be synthesised from the outputs of the integrators (namely the various elements of $\boldsymbol{x}$) and the prescribed elements of $\boldsymbol{u}$, and must, moreover, be made time-dependent according to the form of the various f_r. This process of synthesising the f_r functions is clearly greatly simplified if the f_r are autonomous and linear functions of both $\boldsymbol{x}$ and $\boldsymbol{u}$; indeed, in eqn. 1.5, $\dot{x}_r = f_r = \sum_{s=0}^{n} a_{rs} x_s + \sum_{q=0}^{m} b_{rq} u_q$, so that the input to any integrator is then merely a linear combination of the state-vector and input-vector elements, whcih may readily be synthesised from these elements by amplifiers or attenuators combined with an adder. This situation is represented by Fig. 1.2.

The purpose of simulation is either to compare the output of the simulation with the output of the plant (for the same input) and thus to justify—one hopes—the reasonable accuracy of the mathematical model, or, if one is satisfied with this accuracy, to carry out various forms of tests on the simulation, rather than on the plant, possibly so as not to put the plant out of commission or, in any case, to economise on the cost of such tests. The first of these objectives is the basis of the topic of plant identification, which is outside the scope of this book.

1.2.3 The output equation

Although the output elements are normally, even in a nonlinear, non-autonomous system, constant linear combinations of the state elements, we may postulate a general relation, say

$$\boldsymbol{v} = \boldsymbol{h}(\boldsymbol{x}, \boldsymbol{u}, t)$$

and, if necessary, linearise this equation in the same way as the state equation by writing $\boldsymbol{v}_d = \boldsymbol{h}(\boldsymbol{x}_d, \boldsymbol{u}_d, t)$, using the same datum values $\boldsymbol{x}_d$, $\boldsymbol{u}_d$, as in the state equation. Then also $\boldsymbol{v}_d + \boldsymbol{v}_i = \boldsymbol{h}(\boldsymbol{x}_d + \boldsymbol{x}_i, \boldsymbol{u}_d + \boldsymbol{u}_i, t)$ and by subtraction, using as before a truncated Taylor series, we deduce

$$\boldsymbol{v}_i = \boldsymbol{C}\boldsymbol{x}_i + \boldsymbol{D}\boldsymbol{u}_i$$

where

$$\boldsymbol{C} = \boldsymbol{J}(\boldsymbol{h}:\boldsymbol{x}), \boldsymbol{D} = \boldsymbol{J}(\boldsymbol{h}:\boldsymbol{u}), \text{ both evaluated at } \boldsymbol{x} = \boldsymbol{x}_d, \boldsymbol{u} = \boldsymbol{u}_d.$$

Dropping the suffix as before, the incremental values satisfy

$$\boldsymbol{v} = \boldsymbol{C}\boldsymbol{x} + \boldsymbol{D}\boldsymbol{u} \tag{1.7}$$

which is the canonical linear form of the output equation. In most systems the output equation will be linear in the first place and linearisation is then of course unnecessary.

1.2.3.1 The term Du

The presence of this term in the output equation implies a *direct* contribution of the input to the output (as opposed to an indirect contribution through the medium of the state vector, of which the value is of course affected by the input through the state equation). Whereas this direct contribution is often absent in the physical system, it is sometimes created in the representation in order to eliminate time-derivatives of the input. Consider a simple, linear, electrical circuit consisting of a capacitance K and a resistance R in parallel, this parallel combination being in series with a resistance r; suppose the circuit is excited by an applied voltage $u(t)$ and that the output $v(t)$ is the voltage developed across r. (Fig. 1.3). The current through r may be equated to the current through the K–R parallel combination to give

$$\frac{v}{r} = \frac{u-v}{R} + K(\dot{u} - \dot{v})$$

the single-state equation for the system. If we write $v = x (C = 1, D = 0)$

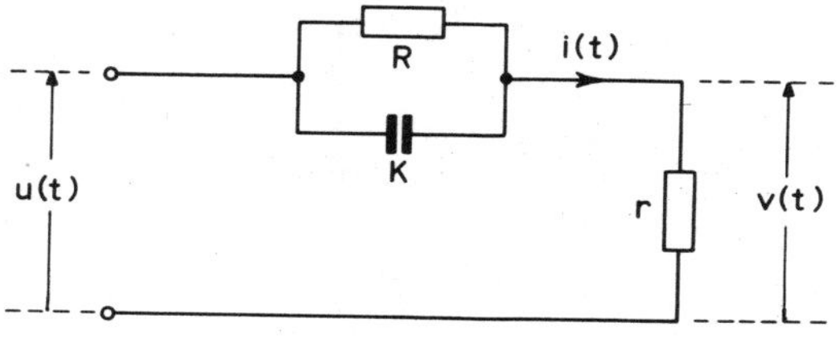

Fig. 1.3

we deduce $\dot{x} = -\dfrac{x}{K}\left(\dfrac{1}{r} + \dfrac{1}{R}\right) + \dfrac{u}{RK} + \dot{u}$ which is *not* of the canonical state equation form since $\dot{u}$ is present. To remove $\dot{u}$ from the original equation it is necessary to write $v = x + u$ $(C = D = 1)$, giving $\dot{x} = -\dfrac{x}{K}\left(\dfrac{1}{r} + \dfrac{1}{R}\right) - \dfrac{u}{rK}$, which *is* of the canonical form.

1.3 The linear equations in discrete time notation

Instead of expressing the state equations as differential equations, it is often necessary (for computer operations or because of the nature of the system) to express these equations in *discrete time* notation, the implication of this phrase being that the quantities in the system are only specified at discrete, usually equispaced, instants of time and not continuously.

We consider system quantities as defined only at $t = kh$, where k is any integer, and denote $\boldsymbol{x}(t), \boldsymbol{u}(t), \ldots$ at $t = kh$ by $\boldsymbol{x}_k, \boldsymbol{u}_k, \ldots$. The interval h is called the sampling interval. We may approximate to $\dot{\boldsymbol{x}}$, admittedly crudely, by $\dot{\boldsymbol{x}} = (\boldsymbol{x}_{k+1} - \boldsymbol{x}_k)/h$, but the approximation may be made as close as we please by making h small enough, if indeed h is arbitrary and not dictated by a physical sampling process in the system. The state equation, eqn. 1.5 then becomes

$$\left.\begin{aligned} \boldsymbol{x}_{k+1} &= \boldsymbol{x}_k + h\boldsymbol{A}\boldsymbol{x}_k + h\boldsymbol{B}\boldsymbol{u}_k \\ &\equiv \boldsymbol{F}\boldsymbol{x}_k + \boldsymbol{G}\boldsymbol{u}_k \end{aligned}\right\} \qquad (1.8)$$

where

$$\boldsymbol{F} = \boldsymbol{I} + h\boldsymbol{A}, \quad \boldsymbol{G} = h\boldsymbol{B}$$

This is the *difference equation* in discrete time notation which replaces the differential equation in continuous time. It is still valid–in a non-autonomous system where $\boldsymbol{A}$ and $\boldsymbol{B}$ may be time-dependent–provided that $\boldsymbol{A}$ and $\boldsymbol{B}$ are replaced by $\boldsymbol{A}_k$ and $\boldsymbol{B}_k$ respectively.

If h is not freely adjustable but is dictated by physical sampling processes in the system (e.g. a radar aerial signal) it may be necessary, in order to obtain sufficient accuracy in calculations, to use better approximations for $\dot{\boldsymbol{x}}_k$, but these will involve a larger number of sampled values of $\boldsymbol{x}$. If h is not so constrained, it is usually preferable to use the cruder approximation with sufficiently small value of h.

The output equation, eqn. 1.7 naturally becomes

$$\boldsymbol{v}_k = \boldsymbol{C}\boldsymbol{x}_k + \boldsymbol{D}\boldsymbol{u}_k \qquad (1.9)$$

1.4 Laplace transformation

This is only practicable for autonomous systems, when A and B are constant. If this is not the case, we have to find the transforms of products of time functions such as $A(t)x(t)$, a process which leads to cumbersome if not intractable convolution integrals. With A and B constant, taking Laplace transforms of eqn. 1.5 gives:

$$\begin{aligned} sX(s) - x_0 &= AX(s) + BU(s) \\ \text{or} \quad (sI - A)X(s) &= x_0 + BU(s) \end{aligned} \tag{1.10}$$

where $X(s)$ is the Laplace transform of $x(t)$ etc., and $x(0) \equiv x_0$. The output equation is again unchanged in form:

$$V(s) = CX(s) + DU(s) \tag{1.11}$$

1.5 Canonical diagonal form of A

If the state vector x is transformed to a new state vector y by the non-singular transformation $x = Ty$, so that $y = T^{-1}x$, the state equation, eqn. 1.5, becomes

$$T\dot{y} = ATy + Bu \quad \text{or} \quad \dot{y} = T^{-1}AT.y + T^{-1}B.u$$

Note that the transformation of A to $T^{-1}AT$ is a *similarity* transformation which leaves the eigenvalues unchanged.

If we suppose that A has *n distinct* eigenvalues $\lambda_1, \lambda_2, \ldots, \lambda_n$ with associated eigenvectors $e_1, e_2, \ldots, e_n$, then, by definition of an eigenvector, $Ae_r = \lambda_r e_r (r = 1, 2, \ldots, n)$, a set of column vector equations which are conveniently written as a single matrix equation

$$AE = E \operatorname{diag} \lambda_r \equiv E\Lambda \quad \text{or} \quad \Lambda = E^{-1}AE$$

where E is the *eigenvector-assembly matrix* of A, with columns $e_1, e_2, \ldots, e_n$, and $\Lambda = \operatorname{diag} \lambda_r$, the diagonal matrix of the eigenvalues. Hence with $T = E$ we obtain $x = Ey, y = E^{-1}.x$ and

$$\dot{y} = \Lambda.y + E^{-1}B.u = \Lambda.y + \beta.u \qquad (\beta \equiv E^{-1}B) \tag{1.12}$$

$$v = CE.y + D.u = \Upsilon.y + D.u \qquad (\Upsilon \equiv C.E) \tag{1.13}$$

in which A is replaced by the diagonal matrix Λ. A number of points deserve comment.

1.5.1 Comments

(i) The transformation is possible even if A is a function of t. In this case, however, the eigenvalues and the eigenvectors may also be functions of t and the resulting form of $\Lambda(t)$ is not so useful for solving the state equation. In addition the matrix CE in the output equation becomes time-dependent even if C is constant.

(ii) The transformation assists the form of the solution of the state equation since the equation for $\dot{y}_r$ contains no other element of y than y_r. Thus if u is given—a necessary condition for any solution—we have a set of n *separate* first-order equations, *one for each* y_r, instead of a set of n *simultaneous* equations for the n elements of x.

(iii) The transformation is still valid even if A has multiple eigenvalues, provided that every such multiple eigenvalue is associated with a corresponding number of linearly independent (l.i.) eigenvectors. A is then *diagonalisable* by the transformation given above. Otherwise, the matrix Λ, though still containing the eigenvalues, repeated if multiple, on its leading diagonal, also has some nonzero elements of value unity on the next upper diagonal, the number of these elements being equal to the difference between n, the order of A, and the total number of l.i. eigenvectors: this is standard matrix theory. The point we wish to stress is that if A should have multiple eigenvalues, this is a 'knife-edge' situation, in the sense that an infinitesimal change in one or other element of A will make the eigenvalues distinct. Coupling this with the fact that A is already an approximation to the system considered, through linearisation procedure, and with the further fact that many theorems in matrix analysis are greatly simplified if a matrix has distinct eigenvalues, we feel justified in making this assumption whenever desirable.

(iv) The transformation of A into Λ has of course its counterparts in the discrete time domain and the Laplace transform domain. In the former it is clear that $F = I + hA$ becomes transformed to $I + h\Lambda$, while in the latter $(sI - A)$ is changed to $(sI - \Lambda)$, both these transformed matrices being diagonal if Λ is diagonal.

1.6 'Phase-variable' canonical form of A

In eqn. 1.12 transform y to z by the nonsingular transformation $z = P.y$ or $y = P^{-1}.z$. We deduce

$$\dot{z} = P\Lambda P^{-1}.z + P\beta.u$$

The phase-variable canonical form of A is defined as the matrix

$$S \equiv \begin{bmatrix} 0 & 1 & 0 & \cdots & 0 & 0 \\ 0 & 0 & 1 & \cdots & 0 & 0 \\ \cdot & \cdot & \cdot & \cdots & \cdot & \cdot \\ 0 & 0 & 0 & \cdots & 0 & 1 \\ -a_n & -a_{n-1} & \cdots & \cdots & -a_2 & -a_1 \end{bmatrix}$$

so that

$$s_{r,r+1} = 1, \quad \{r = 1, 2, \ldots, (n-1)\}$$

$$s_{n,r} = -a_{n+1-r}, \quad (r = 1, 2, \ldots, n)$$

all other elements being zero.

The $a_r (1 \leqslant r \leqslant n)$ in the last row are determined by the fact that, when expanded, $\det(\lambda I - A) \equiv \lambda^n + a_1\lambda^{n-1} + a_2\lambda^{n-2} + \ldots + a_{n-1}\lambda + a_n$, which is therefore the characteristic function of A. It is easily shown that this is also the expansion of $\det(\lambda I - S)$, so that A and S have the same eigenvalues. This suggests that S and A (and therefore also Λ) may be *similar* matrices and that it may be possible to find P such that $S = P\Lambda P^{-1} = PE^{-1}.A.EP^{-1}$.

Suppose A is diagonalisable, so that $\Lambda = \text{diag}\,\lambda_r$, whether or not the λ_r are distinct. If $S = P\Lambda P^{-1}$, then $SP = P\Lambda$. Equate the rth columns:

$$[p_{2r}, p_{3r}, \ldots, p_{nr}, -(a_n p_{1r} + a_{n-1} p_{2r} + \ldots + a_1 p_{nr})]' = \lambda_r p_r$$

where p_r is the rth column of P.

Hence $p_{2r} = \lambda_r p_{1r}$; $p_{3r} = \lambda_r p_{2r} = \lambda_r^2 p_{1r}$; $\ldots$; $p_{nr} = \lambda_r^{n-1} p_{1r}$; and for the last element, after substituting these values of $p_{2r}, p_{3r}, \ldots, p_{nr}$, we obtain

$$(a_n + a_{n-1}\lambda_r + a_{n-2}\lambda_r^2 + \ldots + a_1\lambda_r^{n-1} + \lambda_r^n)p_{1r} = 0,$$

which is satisfied for any value of p_{1r} and for every value of r, since every λ_r satisfies the characteristic equation of A. Hence, for arbitrary nonzero values of p_{1r} (nonzero because P must be nonsingular) we deduce

$$p_r = [1, \lambda_r, \lambda_r^2, \ldots, \lambda_r^{n-1}]'.p_{1r} \; (r = 1, 2, \ldots, n)$$

Hence

$$P = \begin{bmatrix} 1 & 1 & 1 & \cdots & 1 \\ \lambda_1 & \lambda_2 & \lambda_3 & \cdots & \lambda_n \\ \lambda_1^2 & \lambda_2^2 & \lambda_3^2 & \cdots & \lambda_n^2 \\ \cdot & \cdot & \cdot & \cdots & \cdot \\ \lambda_1^{n-1} & \lambda_2^{n-1} & \lambda_3^{n-1} & \cdots & \lambda_n^{n-1} \end{bmatrix} . \text{diag}\, p_{1r} \equiv V. \text{diag}\, p_{1r}$$

where V is known as a Vandermonde matrix and is easily shown to be singular only if the λ_r are *not* distinct. Hence *if A has distinct* *eigenvalues*, $\boldsymbol{P}$ can always be found in an infinite number of ways since the p_{1r} are arbitrary though nonzero. (It may be shown that if $\boldsymbol{A}$ has multiple eigenvalues, then $\boldsymbol{P}$ can still be found, provided that no multiple eigenvalue is associated with more than one l.i. eigenvector, but $\boldsymbol{P}$ in such a case will not be of the form given above, nor will $\boldsymbol{\Lambda}$ be diagonal.)

This canonical form arises naturally if the method given in the Appendix, for expressing a higher order differential equation as a set of first order equations, is followed.

1.7 A note on rank

The rank of $\boldsymbol{B}$ in the state equation, eqn. 1.5 can always be raised to its maximal value of m. For if rank $\boldsymbol{B} < m$, at least one of the columns of $\boldsymbol{B}$, say the rth, $\boldsymbol{b}_r$, is a linear combination of the others, say

$$\boldsymbol{b}_r = \sum_s k_{rs}\boldsymbol{b}_s \qquad (s = 1, 2, \ldots, m \text{ excluding } r)$$

Thus

$$\boldsymbol{B}\boldsymbol{u} = \sum_{q=1}^{m} u_q \,.\, \boldsymbol{b}_q = \sum_s k_{rs}u_r \,.\, \boldsymbol{b}_s + \sum_s u_s \,.\, \boldsymbol{b}_s$$

$$(s = 1, 2, \ldots, m \text{ excluding } r)$$

$$= \sum_s (k_{rs}u_r + u_s) \,.\, \boldsymbol{b}_s$$

Thus $\boldsymbol{B} \,.\, \boldsymbol{u}$ can be expressed as the product of a modified $\boldsymbol{B}$-matrix (in which the rth column of $\boldsymbol{B}$ is deleted) and a modified input-vector of only $(m-1)$ elements of the form $(k_{rs}u_r + u_s)$. The process may, if necessary, be repeated until the modified $\boldsymbol{B}$-matrix has l.i. columns and is therefore of maximal rank.

By a similar argument, the rank of $\boldsymbol{C}$ in the output equation, eqn. 1.6 may also be assumed to be its maximal value p. For if rank $\boldsymbol{C} < p$, at least one of the rows of $\boldsymbol{C}$ is a linear combination of the others, and it then follows that the corresponding output element is either a linear combination of the others or differs from such a combination by some linear combination of the elements of $\boldsymbol{u}$ (caused by the term $\boldsymbol{D}\boldsymbol{u}$ in the output equation). In either case such an output may be considered redundant and omitted, the value of p being thereby decreased by unity: the process is repeatable until the rows of the modified $\boldsymbol{C}$ are l.i., when it is of maximal rank.

We can always therefore modify the representation of the linearised system so that rank $\boldsymbol{B}$, and therefore also rank $\boldsymbol{\beta}(= \boldsymbol{E}^{-1}\boldsymbol{B})$ and rank

Chapter 2

The solution of the linear state equations

2.1 Introduction

By the solution of the state equation is meant here the solution of the following problem (in continuous time): given $\boldsymbol{x}(t)$ at $t = t_0$ and given $\boldsymbol{u}(t)$ for all t in $t_0 \leqslant t \leqslant t_1$, find $\boldsymbol{x}(t)$ at $t = t_1$, t_1 being arbitrary. The process of solution in discrete time or in the Laplace domain has obvious analogous interpretations.

Note that in this formulation of the problem, the boundary conditions are given as the values of *all* the elements of $\boldsymbol{x}(t)$ at one and the same time t_0: if some of these values were given at one instant and some at another instant, the solution would be much more difficult; the problem is then called a 2-point boundary problem.

Note also that if $\boldsymbol{x}(t_1)$ can be found, substitution of its value in the output equation at once yields $\boldsymbol{v}(t_1)$.

2.1.1 The exponential function of a square matrix

This function is defined, as for a scalar quantity, by the exponential series:

$$\exp M = I + M + M^2/2! + M^3/3! + \ldots \textit{ad inf.}$$

the first term being unity matrix of the same order as M. The series may be shown to converge for any M with finite elements. We deduce

$$\exp \boldsymbol{M}t = \boldsymbol{I} + \boldsymbol{M}t + \boldsymbol{M}^2t^2/2! + \boldsymbol{M}^3t^3/3! + \ldots$$

$$\text{Hence } \frac{d}{dt}(\exp \boldsymbol{M}t) = \boldsymbol{M} + \boldsymbol{M}^2t + \boldsymbol{M}^3t^2/2! + \ldots$$

$$= \boldsymbol{M}.\exp \boldsymbol{M}t = (\exp \boldsymbol{M}t).\boldsymbol{M}$$

provided that M is independent of t, just as though M were a scalar. But this parallelism with a scalar M has its limits: e.g. although $\exp(M+N) = \exp(N+M)$, neither of these quantities is *in general* equal to either $(\exp M).(\exp N)$ or $(\exp N).(\exp M)$, these two products in any case having different values. (The four functions are only equal if M and N commute under multiplication, i.e. $MN = NM$).

2.2 Solution of autonomous equation in continuous time

We have eqn. 1.5:

$$\dot{x} - Ax = Bu \text{ with } x(t_0) \equiv x_0 \text{ (given). Find } x(t_1) \equiv x_1.$$

As though dealing with a scalar equation, multiply the equation by the integrating factor $\exp(-At)$, but in this case *pre*multiply:

$$\exp(-At).\dot{x} - \exp(-At).Ax = \exp(-At).B.u(t)$$

or

$$\frac{d}{dt}\{\exp(-At).x\} = \exp(-At).B.u(t)$$

Integrating this result between $t = t_0$ and $t = t_1$ gives

$$\exp(-At_1).x_1 - \exp(-At_0).x_0 = \int_{t_0}^{t_1} \exp(-At).B.u(t).dt$$

Hence

$$x_1 = \exp A(t_1 - t_0).x_0 + \int_{t_0}^{t_1} \exp A(t_1 - t).B.u(t).dt \qquad (2.1)$$

which is the required solution.

The first term, dependent entirely upon the initial value of x, is the *complementary function* part of the solution; the second term, dependent entirely upon the forcing or input vector $u(t)$, is the *particular integral*.

Since the system is autonomous, there is no loss of generality in making $t_0 = 0$. Changing t_1 to t and t to τ as the variable of integration gives the more often quoted version of the solution:

$$x(t) = \exp At.x_0 + \int_0^t \exp A(t-\tau).B.u(\tau).d\tau \qquad (2.2)$$

If A is time-dependent, it is not possible to solve the equation analytically unless A satisfies certain constraints, e.g. $\dot{A}A = A\dot{A}$.

2.2.1 The diagonal canonical form

By analogy with the solution, eqn. 2.2 of eqn. 1.5, it follows that the solution of eqn. 1.12, namely $\dot{\boldsymbol{y}} = \boldsymbol{\Lambda y} + \boldsymbol{E}^{-1}\boldsymbol{B}.\boldsymbol{u}$ is

$$\boldsymbol{y}(t) = \exp \boldsymbol{\Lambda}\, t.\boldsymbol{y}_0 + \int_0^t \exp \boldsymbol{\Lambda}(t-\tau).\boldsymbol{E}^{-1}\boldsymbol{B}.\boldsymbol{u}(\tau).d\tau \qquad (2.3)$$

Precisely the same result may be obtained from eqn. 2.2 by writing $\boldsymbol{x} = \boldsymbol{E}\boldsymbol{y}$ throughout and noting that since $\boldsymbol{A} = \boldsymbol{E\Lambda E}^{-1}$, it follows from the properties of a similarity transformation that $\exp \boldsymbol{A}t = \boldsymbol{E}.\exp\boldsymbol{\Lambda}t.\boldsymbol{E}^{-1}$. Note, moreover, that if $\boldsymbol{A}$ is diagonalisable so that $\boldsymbol{\Lambda} = \text{diag}\ \lambda_r$, whether the λ_r are distinct or not, then it follows that $\boldsymbol{\Lambda}t = \text{diag}\ \lambda_r t$, $\boldsymbol{\Lambda}^2 t^2 = \text{diag}\ \lambda_r^2 t^2$ etc. On substituting these powers of $\boldsymbol{\Lambda}t$ in the exponential series, we at once deduce the simple expression

$$\exp \boldsymbol{\Lambda}t = \text{diag}\,(\exp \lambda_r t)$$

It then follows, as an important deduction from eqn. 2.2, that if $\boldsymbol{u}(t) = \boldsymbol{0}$ in the time interval 0 to t considered, so that the system is left to itself with the initial condition $\boldsymbol{y} = \boldsymbol{y}_0$, then the rth element of $\boldsymbol{y}$ is given by

$$y_r(t) = \exp \lambda_r t.y_{r0}$$

Thus every element of $\boldsymbol{y}$, when the system is unexcited, will vary exponentially with time. The time functions $\exp \lambda_r t$, $(r = 1, 2, \ldots n)$ are known as the *natural modes* of the system.

Note that if the eigenvalues of $\boldsymbol{A}$ are distinct, there are n distinct natural modes. If $\boldsymbol{A}$ has some multiple eigenvalues but is diagonalisable, the number of distinct natural modes equals the number of distinct eigenvalues. (If $\boldsymbol{A}$ is not diagonalisable – a case that we are not dealing with – additional modes of the form $t.\exp \lambda_r t$, $t^2 \exp \lambda_r t$ etc. may exist if λ_r is multiple, the total number of distinct modes lying between the number of distinct eigenvalues and n, according to the structure of $\boldsymbol{A}$.)

Since, moreover, $\boldsymbol{x} = \boldsymbol{E}\boldsymbol{y}$ and hence $x_r = (r\text{th row of } E).\boldsymbol{y}$, it follows that in the unexcited system under discussion, every element of $\boldsymbol{x}$ is a linear combination of the natural modes.

2.3 Solution in discrete time

The state equations in discrete time eqn. 1.8,

$$\boldsymbol{x}_{k+1} = \boldsymbol{F}\boldsymbol{x}_k + \boldsymbol{G}\boldsymbol{u}_k \text{ for an autonomous system, or}$$

$$x_{k+1} = F_k x_k + G_k u_k \text{ for a nonautonomous system,}$$

are virtually their own solution. For, given the initial value of x at some time kh and the value of u at time kh and at subsequent sampling instants, we may at once compute the value of x at time $(k+1)h$; then, increasing k by unity in the equation, at time $(k+2)h$, and so on. There is little point, even in the simpler case of the autonomous system, in obtaining a general expression for x_{k+K} in terms of x_k and the intermediate values of u, since, computationally, it is easier to work one step at a time and since, in addition, this process is more informative because it gives the intermediate values of x.

However, eqn. 1.8, with $F = I + hA$ and $G = hB$, is based upon a crude approximation to $\dot{x}$ at $t = kh$, namely $(x_{k+1} - x_k)/h$. If h is freely adjustable, the approximation may be made as close as desired by making h small enough: this of course increases the number of steps to cover a given time interval. The usual practice is then to compute the final value of x using successively halved values of h until, to some desired degree of accuracy, the answer is unchanged by halving h. If, however, h is fixed by physical processes in the system, more accurate values for F and G may be obtained by using the continuous time solution, eqn. 2.1.

Assuming an autonomous system with A, B constant, let us further assume that, in the interval $kh \leqslant t \leqslant (k+1)h$, $u(t)$ is constant at the value u_k. Then eqn. 2.1 gives, over this interval,

$$x_{k+1} = \exp Ah.x_k + \int_{kh}^{(k+1)h} \exp A\{(k+1)h - t\}.B.u_k.dt$$

On evaluating the integral, noting that $\exp A(k+1)h, B, u_k$ are constant, we obtain

$$x_{k+1} = \exp Ah.x_k + A^{-1}(\exp Ah - I)Bu_k$$

On comparing this result with eqn. 1.8, we find

$$F = \exp Ah = I + Ah + A^2h^2/2! + \ldots$$

$$G = A^{-1}(\exp Ah - I)B = \{h + h^2A/2! + h^3A^2/3! + \ldots\}B \quad (2.4)$$

The earlier approximation used consists of the first two terms of the series for F and the first term of the series for G. Provided that $\exp Ah$ and A^{-1} can be readily computed, the values of F and G in eqn. 2.4 should yield appreciably increased accuracy for a given value of h. The constant value of u used in the interval may just as justifiably be taken as u_{k+1}; even higher accuracy will probably result if it is taken as the mean of its extreme values, namely as $(u_k + u_{k+1})/2$.

2.4 The transition matrix

Consider a linear, unexcited, but possibly time dependent system represented by $\dot{\boldsymbol{x}}(t) = \boldsymbol{A}(t).\boldsymbol{x}(t)$, with $\boldsymbol{x} = \boldsymbol{x}_0$ at $t = t_0$.

Since the system is time dependent, it seems reasonable to suppose that the value of $\boldsymbol{x}(t)$ will depend both on the time t_0 at which the process starts and the time interval $(t - t_0)$ during which it lasts. Suppose then

$$\boldsymbol{x}(t) = \boldsymbol{\Phi}(t - t_0, t_0).\boldsymbol{x}_0$$

The function $\boldsymbol{\Phi}$ is called the *transition matrix* for the system. We deduce

$$\dot{\boldsymbol{\Phi}}.\boldsymbol{x}_0 = \dot{\boldsymbol{x}} = \boldsymbol{A}.\boldsymbol{x} = \boldsymbol{A}.\boldsymbol{\Phi}.\boldsymbol{x}_0$$

and if this equation is to be valid for any given $\boldsymbol{x}_0$, then $\boldsymbol{\Phi}$ must satisfy

$$\dot{\boldsymbol{\Phi}} = \boldsymbol{A}.\boldsymbol{\Phi}$$

Moreover, with $t = t_0, \boldsymbol{x}_0 = \boldsymbol{\Phi}(0, t_0).\boldsymbol{x}_0$, so that we have the boundary condition

$$\boldsymbol{\Phi}(0, t_0) = \boldsymbol{I}_n.$$

Hence the transition matrix may be computed, though not in general found analytically. In the particular case where $\boldsymbol{A}(t) = \boldsymbol{A}$, a constant matrix, however, the differential equation for $\boldsymbol{\Phi}$ may be written

$$\exp(-\boldsymbol{A}t).\dot{\boldsymbol{\Phi}} - \exp(-\boldsymbol{A}t).\boldsymbol{A}\boldsymbol{\Phi} = \boldsymbol{0}$$

i.e. $\dfrac{d}{dt}\{\exp(-\boldsymbol{A}t).\boldsymbol{\Phi}\} = \boldsymbol{0}$, i.e. $\exp(-\boldsymbol{A}t).\boldsymbol{\Phi} = \text{constant} = \exp(-\boldsymbol{A}t_0)$

since $\boldsymbol{\Phi} = \boldsymbol{I}$ when $t = t_0$. Hence in this case $\boldsymbol{\Phi} = \exp \boldsymbol{A}(t - t_0)$ and is therefore only a function of the process time and not explicitly of t_0. Taking, therefore, without any loss of generality $t_0 = 0$, we find $\boldsymbol{\Phi}(t) = \exp \boldsymbol{A}t$, and hence $\boldsymbol{x}(t) = \exp \boldsymbol{A}t.\boldsymbol{x}_0$, which is the same result as that given by eqn. 2.2 in the unexcited case $\boldsymbol{u} = \boldsymbol{0}$. Because of the importance of the function $\exp \boldsymbol{A}t$ we shall explore a number of ways of evaluating it.

(i) By summing the exponential series. This is not recommended. No computer can sum an infinite series and in any case the series may only *start* converging after a very large number of terms. Moreover the computation has to be repeated for every value of t required, though one may of course use relations of the form $\exp(k\boldsymbol{A}t) = (\exp \boldsymbol{A}t)^k$ to curtail computation.

(ii) By using the Sylvester expansion theorem of matrix theory. This gives, provided that the eigenvalues of $\boldsymbol{A}$ are distinct,

$$\exp At = \sum_{r=1}^{n} \exp \lambda_r t \prod_{s \neq r} \frac{A - \lambda_s I}{\lambda_r - \lambda_s} \tag{2.5}$$

The summation now consists of only n terms, each term being the product of a normal mode, $\exp \lambda_r t$, and its matrix coefficient, which is clearly a polynomial in A of degree $(n - 1)$. The evaluation requires, as a preliminary, finding the eigenvalues of A.

(iii) By using similarity transformation principles. We have

$$\Phi(t) = \exp At = E.\exp \Lambda t.E^{-1} = E.\text{diag}(\exp \lambda_r t).E^{-1}$$

provided that A is diagonalisable. The evaluation now consists of one matrix triple product, but requires as preliminaries not merely the eigenvalues of A but their associated eigenvectors. The coefficient of a particular normal mode $\exp \lambda_r t$ is, in this form, the matrix (rth column of E).(rth row of E^{-1}).

(iv) This last method assumes the availability of efficient computer routines for the solution of linear, first-order differential equations with constant coefficients. If in eqn. 2.2 we set $u(t) = 0$, then $x(t) = \Phi(t).x_0$, i.e.

$$x_r(t) = \phi_{r1}(t).x_{10} + \phi_{r2}(t).x_{20} + \ldots + \phi_{rn}(t).x_{n0},\ r = 1, 2, \ldots n.$$

It follows that any element $\phi_{rs}(t)$ of the transition matrix is the value of $x_r(t)$ when all elements of x_0 are made zero, except x_{s0} which is made unity. Hence, by obtaining a computerised solution of the equations $\dot{x} = Ax$ with these initial conditions, the solution for $x(t)$ is precisely the sth column of $\Phi(t)$. The solution is then repeated for all possible values of s, thus giving $\Phi(t)$ for some predetermined value of t.

2.5 Solution in the complex frequency domain

Here again the state equation, eqn. 1.10, namely

$$(sI - A)X(s) = x_0 + BU(s)$$

is virtually its own solution. Premultiplying by $(sI - A)^{-1}$ gives

$$X(s) = (sI - A)^{-1}x_0 + (sI - A)^{-1}BU(s) \tag{2.6}$$

from which $x(t)$ may be obtained by an inverse Laplace transformation.

2.5.1 Properties of $(sI - A)^{-1}$

Note that $(sI - A)^{-1} = \dfrac{\text{adj}\,(sI - A)}{\det\,(sI - A)}$; that adj $(sI - A)$ is a square matrix of which the elements are polynomials in s, of degree $(n - 1)$ for the leading diagonal elements, but of degree $(n-2)$ or less for the other elements; that det $(s\boldsymbol{I}-\boldsymbol{A})$, on the other hand, is a scalar polynomial in s of degree n, identical, apart from s replacing λ, with the characteristic polynomial of $\boldsymbol{A}$, det $(\lambda\boldsymbol{I}-\boldsymbol{A})$, of which the zeros are the eigenvalues λ_r of $\boldsymbol{A}$, distinct or not. It follows that

$$\det(s\boldsymbol{I}-\boldsymbol{A}) = \prod_{r=1}^{n}(s-\lambda_r) \qquad (2.7)$$

It may be proved that, if the eigenvalues are distinct, none of these factors of det $(s\boldsymbol{I}-\boldsymbol{A})$ can be a factor of adj $(s\boldsymbol{I}-\boldsymbol{A})$. If, however, an eigenvalue λ_r has a multiplicity m_r so that the determinant contains the factor $(s-\lambda_r)^{m_r}$, then some *lower* power of $(s-\lambda_r)$ *may* be a factor of the adjoint.

On comparing the solution of eqn. 2.2, taking $\boldsymbol{u}(t)=\boldsymbol{0}$, with the solution of eqn. 2.6, with $\boldsymbol{U}(s)=\boldsymbol{0}$, we have on the one hand $\boldsymbol{x}(t)=\boldsymbol{\Phi}(t)\boldsymbol{x}_0$ and on the other $\boldsymbol{X}(s)=(s\boldsymbol{I}-\boldsymbol{A})^{-1}\boldsymbol{x}_0$, whatever $\boldsymbol{x}_0$ may be. We deduce the important relationship:

$$\text{Laplace transform of } \boldsymbol{\Phi}(t)\text{, i.e. of } \exp \boldsymbol{A}t = (s\boldsymbol{I}-\boldsymbol{A})^{-1} \qquad (2.8)$$

This equality may also be established more generally by using the binomial expansion

$$(s\boldsymbol{I}-\boldsymbol{A})^{-1} = \frac{1}{s}\left\{\boldsymbol{I}+\frac{\boldsymbol{A}}{s}+\frac{\boldsymbol{A}^2}{s^2}+\ldots\right\}$$

(assuming s to be large enough for the series to converge). The inverse Laplace transform of the series, term by term, is found to be the exponential series for exp $\boldsymbol{A}t$.

This result may be used to establish the consistency of the input-dependent terms in the solutions of eqns. 2.2 and 2.6. For, by the convolution theorem of Laplace transformation,

$$\mathcal{L}^{-1}\{(s\boldsymbol{I}-\boldsymbol{A})^{-1}.\boldsymbol{B}\boldsymbol{U}(s)\} = \text{convolution integral of } \boldsymbol{\Phi}(t) \text{ and } \boldsymbol{B}\boldsymbol{u}(t)$$

$$= \int_0^t \boldsymbol{\Phi}(t-\tau).\boldsymbol{B}\boldsymbol{u}(\tau)\,d\tau$$

which is precisely the input-dependent term in eqn. 2.2.

2.5.2 The transfer matrix

This is defined as that matrix, of order $p \times m$, which, when premultiplying the input-vector $U(s)$, gives the output vector $V(s)$, *on the supposition that the state vector* $\boldsymbol{x}(t)$ *is initially zero*, i.e. $\boldsymbol{x}_0 = \boldsymbol{0}$. With this condition, eqn. 2.6 simplifies to

$$\boldsymbol{X}(s) = (s\boldsymbol{I}-\boldsymbol{A})^{-1}\boldsymbol{B}\boldsymbol{U}(s)$$

which, when substituted in the output equation, eqn. 1.11, gives

$$\boldsymbol{V}(s) = \{\boldsymbol{C}(s\boldsymbol{I}-\boldsymbol{A})^{-1}\boldsymbol{B}+\boldsymbol{D}\}\,\boldsymbol{U}(s)$$

Hence the value of the transfer matrix, denoted here by $\boldsymbol{G}(s)$, is

$$\boldsymbol{G}(s) = \boldsymbol{C}(s\boldsymbol{I}-\boldsymbol{A})^{-1}\boldsymbol{B}+\boldsymbol{D} \qquad (2.9)$$

The transfer matrix is therefore uniquely determined by the four matrices $\boldsymbol{A}, \boldsymbol{B}, \boldsymbol{C}, \boldsymbol{D}$, which specify the system equations. As will be seen in the next Chapter, however, the transfer matrix does *not* specify the four system matrices, nor, indeed, does it even specify the order of the state vector!

2.6 Examples 2

1. A simple system is described by the differential equation

$$\frac{d^2\theta}{dt^2} + 3\frac{d\theta}{dt} + 2\theta = 2u$$

Write this equation in state vector form with $x_1 = \theta, x_2 = \dot{x}_1$.

Evaluate the transition matrix $\exp(\boldsymbol{A}t)$: (*a*) by evaluating $(s\boldsymbol{I}-\boldsymbol{A})^{-1}$ and finding its inverse Laplace transform; (*b*) by using the Sylvester expansion theorem.

If, in the interval $0 \leqslant t \leqslant 1$, $u = t$, and if $\boldsymbol{x}(0) = \begin{bmatrix} -1 \\ -1 \end{bmatrix}$, find $\boldsymbol{x}(1)$ correct to four significant figures in each element.

Using discrete time notation, with $h = \frac{1}{2}, \frac{1}{4}$ successively, repeat the calculation of $\boldsymbol{x}(1)$:

(i) with $\boldsymbol{F} = \boldsymbol{I} + h\boldsymbol{A}$, $\boldsymbol{G} = h\boldsymbol{B}$; (ii) with $\boldsymbol{F} = \exp(\boldsymbol{A}h)$, $\boldsymbol{G} = \boldsymbol{A}^{-1}\{\exp(\boldsymbol{A}h) - \boldsymbol{I}\}\boldsymbol{B}$; (iii) as in (ii) but taking the constant value of u in a sampling interval as its mean value over that interval instead of its initial value.

Compare the various results obtained and draw deductions.

2. Return to Example 1.3 and use the linearised state equation obtained for that problem with $r = R, \dot{\theta} = \Omega, \boldsymbol{u} = \boldsymbol{0}$, as datum values.

At $t = 0$, u_1 is suddenly increased to the value U for a short time T, so that effectively u_1 is an impulse function $UT\delta(t)$. Estimate the subsequent behaviour with time of $r, \dot{r}$ and $\dot{\theta}$ and deduce the equation of the new orbit. The new orbit would be expected to be an ellipse: is it and if not, why not?

Repeat the example with u_2 replacing u_1.

3. The linearised form of the equations describing a turbo-prop engine control system is, in the complex frequency domain:

$$X(s) = \frac{1}{1+s}\begin{bmatrix} -2 & 0 \\ 4 & 0 \\ 0 & 3 \\ 0 & 1+4s \end{bmatrix} U(s) \quad \text{and} \quad V(s) = \begin{bmatrix} 1 & 0 & 1 & 0 \\ 0 & 1 & 0 & 1 \end{bmatrix} X(s)$$

in which u_1, u_2, v_1, v_2 are deviations in propeller blade angle, fuel rate, engine speed and turbine inlet temperature respectively.

If (i) u_1, (ii) u_2, is a unit step function at $t = 0$, find $v(t)$ subsequently. If, in each case, the step-function is cut off at $t = 2$, find $v(t)$ for $t \geqslant 2$.

By writing the system equations in the time domain obtain a possible set of $\boldsymbol{A}, \boldsymbol{B}, \boldsymbol{C}, \boldsymbol{D}$ matrices and find (*a*) the eigenvalues of $\boldsymbol{A}$, (*b*) the transition matrix for the system.

Chapter 3

Controllability, observability and transfer-matrix representation

3.1 Controllability

A system is said to be (completely) state-controllable if, for any t_0, it is possible to find some finite input-vector $\boldsymbol{u}(t)$ which will transfer *any* given initial state $\boldsymbol{x}_0$ to *any* final state $\boldsymbol{x}_1$ in a finite time interval $(t_1 - t_0)$. An identical definition applies to output-controllability, substituting 'output $\boldsymbol{v}$' for 'state $\boldsymbol{x}$'.

The word 'completely' in the definition seems a little unnecessary: systems which are only partly controllable, i.e. systems for which only some of the state elements can be brought from an arbitrary initial value to an arbitrary final value, are not of importance. Little, if any, theory has been developed regarding controllability and observability of non-linear systems; although the theory may be extended to nonautonomous systems, such systems are rare; it will be assumed in this Chapter that the system representation is therefore linear and autonomous, namely

$$\left.\begin{aligned}\dot{\boldsymbol{x}} &= \boldsymbol{A}\boldsymbol{x} + \boldsymbol{B}\boldsymbol{u}\\ \boldsymbol{v} &= \boldsymbol{C}\boldsymbol{x} + \boldsymbol{D}\boldsymbol{u}\end{aligned}\right\}\text{ in continuous time, or }\left.\begin{aligned}\boldsymbol{x}_{N+1} &= \boldsymbol{F}\boldsymbol{x}_N + \boldsymbol{G}\boldsymbol{u}_N\\ \boldsymbol{v}_N &= \boldsymbol{C}\boldsymbol{x}_N + \boldsymbol{D}\boldsymbol{u}_N\end{aligned}\right\}\text{ in discrete}$$

time.

Since the system is autonomous, moreover, the phrase 'for any t_0' in the definition is superfluous and, for convenience, we shall take $t_0 = 0$.

It is normally assumed that the end-point $\boldsymbol{x}(t_1)$ is the origin of the state-space: this assumption requires justification and is not necessary.

3.1.1 Criterion of state-controllability

From the solution to eqn. 2.2:

$$\boldsymbol{x}_1 = \exp \boldsymbol{A}t_1 . \boldsymbol{x}_0 + \int_0^{t_1} \exp \boldsymbol{A}(t_1 - t).\boldsymbol{B}\boldsymbol{u}(t).dt$$

or

$$\exp(-At_1).x_1 - x_0 = \int_0^{t_1} \exp(-At).Bu(t).dt \qquad (3.1)$$

Since x_0 and x_1 are to be independently arbitrary vectors, the left hand side of eqn. 3.1 and therefore also the right hand side, must be arbitrary vectors. In the integrand, it is known, either from the Cayley-Hamilton theorem or from the Sylvester form, see eqn. 2.5, that $\exp(-At)$ may be expanded as a polynomial of degree $(n-1)$ in A. Write therefore

$$\exp(-At) = \sum_{r=0}^{n-1} k_r(t).A^r$$

Also, the coefficient of the input element $u_s(t)$ in the product $Bu(t)$ is the sth column of B, denoted here by b_s. Hence

$$Bu(t) = \sum_{s=1}^{m} b_s.u_s(t)$$

Making both these substitutions in eqn. 3.1 and rearranging gives:

$$\exp(-At_1).x_1 - x_0 = \sum_{r=0}^{n-1} \sum_{s=1}^{m} A^r b_s \int_0^{t_1} k_r(t) u_s(t)\, dt$$

The various integrals in this expression will have a certain value which may be adjusted by choosing suitable forms for the input elements $u_s(t)$. But each integral is the coefficient of a column matrix $A^r b_s$, namely the sth column of $A^r B$. Hence the right hand side of eqn. 3.1 is some linear combination of the columns of the assembly matrix

$$M \equiv [B, AB, A^2B, \ldots, A^{n-1}B] \text{ of order } n \times nm$$

with coefficients which are arbitrarily adjustable through the u_s. If this linear combination of columns is to represent an arbitrary vector in n-dimensional space, the columns themselves must span this space, i.e.

$$\text{Rank } M = \text{Rank } [B, AB, A^2B, \ldots, A^{n-1}B] = n \qquad (3.2)$$

which is therefore the required criterion of complete state-controllability.

If Rank $M \equiv R < n$, then the right hand side of eqn. 3.1 can only represent a vector in R-dimensional space, which places a constraint on x_0 and x_1.

3.1.1.1 Use of the Gram matrix

If m and n are at all large, it may be a very onerous task to assess the rank of M, though it may save time to note that if, in computing the successive components of M, namely B, AB etc., it is found that adding A^rB to the sequence does not increase the rank of the sequence, then adding $A^{r+1}B, A^{r+2}B, \ldots$ to the sequence will not increase the rank either, so that there is no point in completing the sequence.

Alternatively, it may be quicker to use the theory of the Gram matrix, G, defined for any matrix $K_{m\times n}$ by $G = K^*K$, where K^* is the conjugate transpose of K; if K is real, then $G = K'K$. Its importance here arises from the fact that G is nonsingular if and only if rank $K = n$.

Since M is real and of order $n \times nm$, consider the Gram matrix of M', i.e. $G = MM'$, which is a square matrix of order n. Then M', and therefore M, can only be of rank n if $G = MM'$ is nonsingular, which is an alternative form of the criterion.

3.1.1.2 Diagonal canonical form

It is assumed that A has distinct eigenvalues. With $x = Ey$, the state equation is eqn. 1.12, namely $\dot{y} = \Lambda y + \beta u$, where $\beta = E^{-1}B$. Since E is nonsingular, to every x corresponds one and only one y, and *vice versa*. Thus if $u(t)$ can be found to transfer x from an arbitrary x_0 to an arbitrary x_1, $u(t)$ can also be found to transfer y from an arbitrary y_0 to an arbitrary y_1: state-controllability in the x-space is identical with state-controllability in the y-space. The criterion of eqn. 3.2 may therefore be stated as

$$\text{Rank}\ [\beta, \Lambda\beta, \Lambda^2\beta, \ldots, \Lambda^{n-1}\beta] = n$$

A typical column in this matrix, the sth column of the $(r+1)$th component, is

$$[\lambda_1^r\beta_{1s}, \lambda_2^r\beta_{2s}, \ldots, \lambda_n^r\beta_{ns}]';\ r = 0, 1, \ldots, (n-1);\ s = 1, 2, \ldots, m$$

This column may clearly be written as the product of diag $\beta_{ps}(p = 1, 2, \ldots, n) \equiv D_s$ and the column $[\lambda_1^r, \lambda_2^r, \ldots, \lambda_n^r]$. Keeping s fixed but varying r gives therefore the columns of the product D_sL, say, where

$$L = \begin{bmatrix} 1 & \lambda_1 & \lambda^2 & \ldots \lambda_1^{n-1} \\ 1 & \lambda_2 & \lambda_2^{1/2} & \ldots \lambda_2^{n-1} \\ . & . & . & \ldots . \\ . & . & . & \ldots . \\ 1 & \lambda_n & \lambda_n^2 & \ldots \lambda_n^{n-1} \end{bmatrix}$$

and is nonsingular since the eigenvalues are distinct. (Compare the matrix $V = L'$ in Section 1.6). The controllability criterion therefore becomes

$$\text{rank } [D_1L, D_2L, \ldots, D_mL]_{n \times nm} = n$$

or

$$\text{rank } [D_1, D_2, \ldots, D_m]_{n \times nm} = n$$

since the postmultiplication of each member by the nonsingular matrix L does not alter the rank. Using the Gram matrix criterion, but noting that although B is real, $\boldsymbol{\beta}$ may be complex if A has any complex eigenvalues, we deduce the criterion

$$\det \{[D_1, D_2, \ldots, D_m].[D_1, D_2, \ldots, D_m]^*\} \neq 0$$

i.e.

$$\det [D_1D_1^* + D_2D_2^* + \ldots + D_mD_m^*] \neq 0$$

in which

$$D_sD_s^* = \text{diag} \,|\beta_{ps}|^2, \text{ since } D_s = \text{diag}\, \beta_{ps}$$

It follows that the matrix $[D_1D_1^* + D_2D_2^* + \ldots + D_mD_m^*]$ is a diagonal matrix of which the pth diagonal element is $\{|\beta_{p1}|^2 + |\beta_{p2}|^2 + \ldots + |\beta_{pm}|^2\}$ and hence that its determinant can only vanish if, for some value of p, all β_{ps} vanish. The criterion of state-controllability therefore takes the simple form: *the system is state-controllable if and only if* $\boldsymbol{\beta} = E^{-1}B$ *has no null row.*

An important deduction may be drawn. For if the system is *not* completely controllable and the qth row of $E^{-1}B$, say, is null, then the qth state equation simplifies to $\dot{y}_q = \lambda_q y_q$, with solution $y_q = \exp(\lambda_q t).y_{q0}$. In other words, the natural mode $\exp(\lambda_q t)$, which occurs nowhere else in the state equations, depends entirely for its existence on the initial value y_{q0} and is completely unaffected by the input vector u.

3.1.1.3 Controllability in discrete time

The state equation is now $x_{k+1} = Fx_k + Gu_k$. Starting with $k = 0$, corresponding to $t_0 = 0$, and using the equation iteratively with increasing values of k, we may deduce

$$x_K - F^K x_0 = F^{K-1}Gu_0 + F^{K-2}Gu_1 + \ldots + FGu_{K-2} + Gu_{K-1}$$

If x_0 and x_K are to be arbitrary n-vectors, the right hand side must also be arbitrary. A typical term in the right hand side is $F^rGu_{K-r-1}(r = 0, 1, \ldots, k-1)$ and may be considered as the sum of the products of the several columns of F^rG and the corresponding elements of u_{K-r-1}, over which we have a free choice. Thus, if the right hand side is to represent an arbitrary vector in n-dimensional space, we require

$$\text{Rank } [G, FG, F^2G, \ldots, F^{K-1}G] = n$$

which is therefore the criterion of complete state-controllability.

However, since F is of order n, by the Cayley-Hamilton theorem, any power of $F \geqslant n$, may be expressed as a polynomial in F of degree $(n-1)$. It follows that there is no point in making $K > n$, since we shall merely be adding new columns which are linear combinations of earlier ones. Hence the criterion may be stated as

$$\text{Rank } [G, FG, F^2G, \ldots, F^{n-1}G] = n \tag{3.3}$$

Moreover, if this criterion is satisfied, the transit from x_0 to x_K may be accomplished in, at most, n sampling intervals: indeed if the earlier criterion is satisfied for some value of $K < n$, the transit only requires K sampling intervals.

3.1.1.4 Controllability in the complex frequency domain
The solution given in eqn. 2.6 to the state equation may be written

$$\begin{aligned} X(s) &= \{\text{adj } (sI - A).x_0/\det (sI - A)\} \\ &\quad + \{\text{adj } (sI - A).B.U(s)/\det (sI - A)\} \end{aligned}$$

where, as seen earlier, $\det (sI - A) = \prod_{r=1}^{n} (s - \lambda_r)$. Assuming that the eigenvalues of A are distinct, no factor $(s - \lambda_r)$ of $\det (sI - A)$ can be a factor of adj $(sI - A)$. However, B may be such that some $(s - \lambda_r)$ is a factor of the product adj $(sI - A).B$. If this is the case, such a factor will cancel out in the second term of the expression for $X(s)$ and, on inverse Laplace transformation, it then follows that the associated natural mode, $\exp \lambda_r t$, will only arise in the first term. In other words this mode will, in the various elements of $x(t)$, be dictated entirely by the initial state x_0 and will be unaffected by the input $u(t)$. As seen in Section 3.1.1.2, this situation corresponds to incomplete state-controllability. Hence, *if the system is to be completely state-controllable,* adj $(sI - A).B$ *must have no common factor with* $\det (sI - A)$.

3.1.2 Output controllability

If, as is often the case, $D = 0$, so that the output equation simplifies to $v = C.x$, an analysis similar to that of Section 3.1.1 shows that the system is output-controllable if and only if

$$\text{Rank } [CB, CAB, CA^2B, \ldots, CA^{n-1}B] = p$$

i.e. if and only if Rank $C.M = p$, where M is given by eqn. 3.2. This is only possible if rank C has its maximal value p, a condition which can always be satisfied (see Section 1.7) by omitting redundant output elements. If $\boldsymbol{C}$ is of rank p and $\boldsymbol{M}$ is of maximal rank n, rank $\boldsymbol{C}.\boldsymbol{M} = p$: in other words if rank $\boldsymbol{C} = p$ and if the system is state-controllable, then the system is also output-controllable. It is not, however, *essential* that rank $\boldsymbol{M}$ should equal n though of course rank $\boldsymbol{M}$ must not be smaller than p: it *is* essential that rank $\boldsymbol{C} = p$.

If $\boldsymbol{D} \neq \boldsymbol{0}$, its presence clearly facilitates output-controllability by providing a direct channel by which the input affects the output (in addition to the indirect channel through the state vector). It may be shown that a quite general criterion of output controllability is

$$\text{Rank } [\boldsymbol{CB}, \boldsymbol{CAB}, \ldots, \boldsymbol{CA}^{n-1}\boldsymbol{B}, \boldsymbol{D}] = p \tag{3.4}$$

3.2 Observability

A system is said to be observable in the interval $t_0 \leqslant t \leqslant t_0 + T$ if, given $\boldsymbol{u}(t)$ in this interval, $\boldsymbol{x}(t_0)$ can be determined from a knowledge of (i.e. any number of observations of) $\boldsymbol{v}(t)$ in the interval. Since we are only concerned with autonomous systems, we again assume $t_0 = 0$.

3.2.1 Observability in continuous time

Substituting the state-equation solution of eqn. 2.2 in the output equation eqn. 1.7 gives, after rearrangement:

$$\boldsymbol{v}(t) - \boldsymbol{D}.\boldsymbol{u}(t) - \boldsymbol{C}\int_0^t \exp \boldsymbol{A}(t-\tau).\boldsymbol{B}.\boldsymbol{u}(\tau).d\tau = \boldsymbol{C}.\exp \boldsymbol{A}t.\boldsymbol{x}_0 (0 \leqslant t \leqslant T)$$

In this expression, the second and third terms on the left hand side are directly calculable since $\boldsymbol{A}, \boldsymbol{B}, \boldsymbol{C}, \boldsymbol{D}$ and $\boldsymbol{u}(\tau)$ are known. Hence any observation of $\boldsymbol{v}(t)$ is equivalent to an observation of the left hand side. The problem of observability is therefore unchanged if we omit the second and third terms, which is equivalent to making $\boldsymbol{u} = \boldsymbol{0}$ throughout. We thus reset the problem as:

$$\text{if } \boldsymbol{v}(t) = \boldsymbol{C}.\exp \boldsymbol{A}t.\boldsymbol{x}_0,$$

can $\boldsymbol{x}_0$ be determined by a knowledge of $\boldsymbol{v}(t)$?

Let $\exp \boldsymbol{A}t$ be expressed as a polynomial in $\boldsymbol{A}$ of degree $(n-1)$:

$$\exp \boldsymbol{A}t = K_0(t)\boldsymbol{I}_n + K_1(t)\boldsymbol{A} + K_2(t)\boldsymbol{A}^2 + \ldots + K_{n-1}(t)\boldsymbol{A}^{n-1}$$

where the $K_r(t)$ are scalar coefficients. Then

$$v(t) = [K_0(t)C + K_1(t)CA + K_2(t)CA^2 + \ldots + K_{n-1}(t)CA^{n-1}].x_0$$

$$= [K_0(t)I_p, K_1(t)I_p, \ldots, K_{n-1}(t)I_p].\begin{bmatrix} C \\ CA \\ \cdot \\ \cdot \\ \cdot \\ CA^{n-1} \end{bmatrix} x_0 \text{ in partitioned form.}$$

If we now suppose that observations of $v(t)$ are made at $t = t_s$ ($s = 1, 2, \ldots, N$) then, denoting for brevity $K_r(t_s) \equiv K_{rs}$ and $v(t_s) \equiv v_s$, we obtain:

$$\begin{bmatrix} v_1 \\ v_2 \\ \cdot \\ \cdot \\ v_N \end{bmatrix} = \begin{bmatrix} K_{01}I_p & K_{11}I_p & \ldots K_{(n-1)1}I_p \\ K_{02}I_p & K_{12}I_p & \ldots K_{(n-1)2}I_p \\ \cdot & \ldots & \cdot \\ \cdot & \ldots & \cdot \\ K_{0N}I_p & K_{1N}I_p & \ldots K_{(n-1)N}I_p \end{bmatrix}_{Np \times np} . \begin{bmatrix} C \\ CA \\ \cdot \\ \cdot \\ CA^{n-1} \end{bmatrix}_{np \times n} .x_0 \quad (3.5)$$

If x_0 is to be deducible from these Np scalar equations for its n elements, the coefficient matrix must be of rank n. This requires in the first place as a *necessary* condition that the rank of the second factor (or its transpose) should equal n:

$$\text{rank } [C', A'C', A'^2C', \ldots, A'^{(n-1)}C'] = n \quad (3.6)$$

As far as the first factor is concerned, a *sufficient* (but not a necessary) condition for the product to be of rank n is that the rank of the first factor should be maximal, i.e. equal to np, since we suppose N as large as we please. Now it is easy to show that the rank of this first factor is p times the rank of the matrix $[K_{rs}]$ $\{r = 0, 1, \ldots, (n-1); s = 1, 2, \ldots, N\}$. Since it may be shown that the K_{rs}, i.e. the $K_r(t_s)$, are an l.i. set of functions of t_s, the rank of the matrix $[K_{rs}]$ can always be made equal to its maximum value n by a suitable choice of the various t_s and by making $N \geqslant n$. Thus the required constraint on the first factor can always be met. *There remains therefore only the condition* stated in eqn. 3.6 *as a necessary and sufficient condition of observability*.

This condition is independent of the nature of the eigenvalues of A. If however A has multiple eigenvalues and satisfies a polynomial equation of degree $\nu < n$, then there is no point in extending the compound matrix in eqn. 3.6 beyond the element $A'^{(\nu-1)}C'$, since any further element will be a linear combination of the first ν elements.

3.2.1.1 Diagonal canonical form

With Λ replacing A and $CE = \Upsilon$ replacing C, and noting that if A is diagonalisable, $\Lambda = \Lambda'$, the criterion stated in eqn. 3.6 becomes

$$\text{rank}\ [\Upsilon', \Lambda\Upsilon', \Lambda^2\Upsilon', \ldots, \Lambda^{(n-1)}\Upsilon'] = n$$

With Υ' replacing β, this is identical with the state-controllability criterion of Section 3.1.1.2, which was found to be equivalent, assuming that A has distinct eigenvalues, to β having no null rows. The corresponding criterion of observability is therefore that Υ *shall have no null columns*.

As an important deduction, if the system is *not* observable and therefore Υ *has* a null column, the output equation (putting $u(t) = 0$ as before), namely $v(t) = \Upsilon . y(t)$, shows that all output elements are independent of some element of $y(t)$ and therefore do not contain one (or more) of the natural modes of the system, which are present in the state vector $x(t)$.

3.2.2 In discrete time

Setting $u(t) = 0$ gives, in eqn. 1.8, $x_{k+1} = F . x_k$ which, with the output equation, $v_k = Cx_k$, gives on iteration

$$v_k = C.F^k.x_0$$

$$v_{ik} = c_i.F^k.x_0,\ k = 0, 1, \ldots, k_1 \text{ say, where } k_1 h = t_1,$$
$$i = 1, 2, \ldots, p,$$

and c_i denotes the ith row of C. If the n elements of x_0 are to be obtainable uniquely from these $p(k_1 + 1)$ equations, their rank must be n, i.e. (i) $p(k_1 + 1) \geqslant n$, which sets a lower limit to the number of observational instants required, and (ii) the rank of the rows of C, CF, $\ldots, CF^{k_1} = n$; or, since F^n and higher powers of F can be expressed as linear combinations of $F^0, F, F^2, \ldots, F^{n-1}$, the criterion becomes

$$\text{rank}\ [C', F'C', F'^2C', \ldots, F'^{(n-1)}C'] = n$$

3.2.3 In the complex frequency domain

Again setting $u(t) = 0 = U(s)$, it follows from eqn. 2.6 and the output equation $V(s) = CX(s)$ that

$$V(s) = C.\text{adj}\ (sI - A).x_0/\det\ (sI - A)$$

It was noted in Section 3.2.1 that nonobservability implies the absence of one or more natural modes from the output vector. It follows at once that the criterion of observability is that there should be no cancellation of factors of the form $(s-\lambda_r)$ between $C.\text{adj}\,(sI-A)$ and $\det\,(sI-A)$, assuming as before that A has distinct eigenvalues.

3.3 The duality of state-controllability and observability

For ease of comparison, the criteria of state-controllability (C) and observability (O) obtained in this Chapter are listed below, on the supposition that A has distinct eigenvalues:

Continuous time:	C	rank $[B, AB, A^2B, \ldots, A^{n-1}B] = n$
	O	rank $[C', A'C', A'^2C', \ldots, A'^{(n-1)}C'] = n$
Diagonal form	C	$\beta = E^{-1}B$ must have no null row
	O	$\gamma = CE$ must have no null column
Discrete time	C	rank $[G, FG, F^2G, \ldots, F^{n-1}G] = n$
	O	rank $[C', F'C', F'^2C', \ldots, F'^{(n-1)}C'] = n$
s-domain	C	no common factor between adj $(sI-A).B$ and det $(sI-A)$
	O	no common factor between $C.$adj $(sI-A)$ and det $(sI-A)$

Note first that none of these criteria involves the matrix D. Note next, and this is more important, that if the system represented by the matrices (A, B, C, D) is controllable (or observable) then the system (A', C', B', D) is observable (or controllable). In this sense these two systems are *dual systems* with respect to the properties of state-controllability and observability. Note further that if the orders of the state, input and output vectors in one system are (n, m, p) respectively, then for the dual system they are (n, p, m).

The duality property is obvious in the continuous time criteria. In the diagonal form, since $A = E\Lambda E^{-1}$, $A' = (E')^{-1}\Lambda E'$, so that changing A to A' is equivalent to changing E to $(E')^{-1}$; since also B changes to C', $E^{-1}B$ changes to $E'C' = (CE)'$. In discrete time, $G = hB$ becomes hC', while $F = I + hA$ becomes $I + hA' = F'$. Finally, in the s-domain, adj $(sI-A).B$ becomes adj $(sI-A').C' = [C.\text{adj}\,(sI-A)]'$.

Finally note that noncontrollability of the state implies that some natural mode, normally present in the state vector, is not affected by the input, whereas nonobservability implies that some natural mode, normally present in the state-vector, is absent from the output.

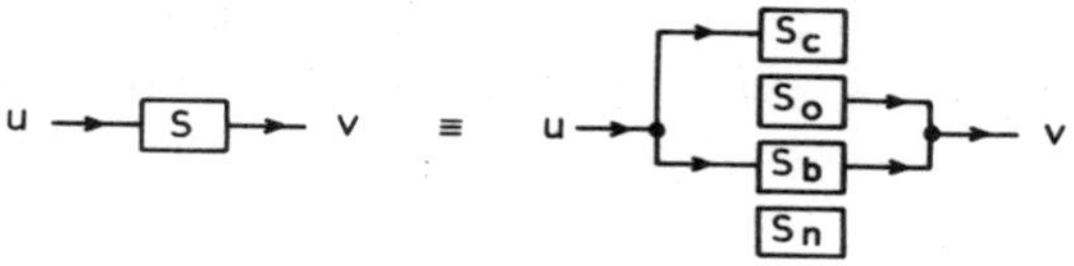

Fig. 3.1 System and equivalent subsystems

3.4 Equivalent subsystems and transfer matrices

In the light of this last statement, it is interesting to divide the natural modes of the state-vector into four groups: group *o*, those modes which are *o*bservable but not controllable; group *c*, those which are *c*ontrollable but not observable; group *b*, those which are *b*oth, and group *n*, those which are *n*either. This concept is pictorialised in Fig. 3.1, showing the system S (within which the state-vector contains in general all the natural modes), relating the input vector $\boldsymbol{u}(t)$ to the output vector $\boldsymbol{v}(t)$, as equivalent to the four subsystems S_o, S_c, S_b and S_n containing only the modes of groups o, c, b, n respectively. The input only affects controllable modes (groups c and b), the output is only contributed to by observable modes (groups o and b).

It will be seen that the only connection between $\boldsymbol{u}(t)$ and $\boldsymbol{v}(t)$ is through the subsystem S_b. Hence S_b is the only subsystem which can affect the transfer matrix of the system S relating $\boldsymbol{V}(s)$ to $\boldsymbol{U}(s)$ and this transfer matrix must in fact also be the transfer matrix of S_b. Hence if the only information we have about the system S is its transfer matrix $\boldsymbol{G}(s)$, then the only subsystem which we can legitimately deduce is S_b; but, if we choose, we can add to this S_b *any* S_c, S_o or S_n without affecting $\boldsymbol{G}(s)$, even though such additions, by increasing the number of the natural modes, increase the value of n. In other words a given $\boldsymbol{G}(s)$ may have any number of representations in $(\boldsymbol{A}, \boldsymbol{B}, \boldsymbol{C}, \boldsymbol{D})$ form, in all of which m and p are of course fixed, but in which n has any value greater than or equal to a certain minimal value, this minimal value being determined by the number of modes present in the equivalent S_b, or, more precisely, by the degree of the denominator of $\boldsymbol{G}(s)$ in eqn. 2.9 after any cancellations have taken place.

In brief, we have three representations of a given physical system to consider. First, if we can analyse the system in detail, breaking down its interdependences to physically present first-order differential equations, we obtain a correct state representation, including any noncontrollable or nonobservable modes which the physical system may contain.

Secondly, either from this representation or by other means, possibly experimental, we may obtain the output-to-input relation, the transfer matrix $G(s)$, or its precise counterpart, a minimal state representation, in which any modes in groups o, c or n are eliminated: if such modes are actually present physically, such a representation is incomplete. Thirdly, the given $G(s)$ may be expressed by a nonminimal state form, in which case the representation contains completely fictitious modes in groups o, c and n and is therefore again an incorrect representation of the physical system. (This third type is clearly to be avoided if possible, since it increases the order of A, B and C while at the same time introducing falsity of representation.)

We give two examples to illustrate some of these points and to introduce possible techniques which may be useful.

3.4.1 Example 1

Given the transfer matrix

$$G(s) = \frac{1}{(s+2)(s+3)(s+4)}\begin{bmatrix} s^2+2s+2 & 2(s+1) \\ -7s^2-24s-24 & 2s(s+1) \end{bmatrix}$$

obtain a minimal state-vector representation. (Note $m = p = 2$)

Probably the quickest way of solving this problem is to argue that the minimal representation must have $n = 3$ and be such that the eigenvalues of A are the zeros of the denominator of $G(s)$, namely $-2, -3$ and -4. In diagonal canonical form, we therefore postulate

$$\mathbf{\Lambda} = \begin{bmatrix} -2 & 0 & 0 \\ 0 & -3 & 0 \\ 0 & 0 & -4 \end{bmatrix}; \quad \boldsymbol{\beta} = \begin{bmatrix} \beta_{11} & \beta_{12} \\ \beta_{21} & \beta_{22} \\ \beta_{31} & \beta_{32} \end{bmatrix};$$

$$\boldsymbol{\gamma} = \begin{bmatrix} \gamma_{11} & \gamma_{12} & \gamma_{13} \\ \gamma_{21} & \gamma_{22} & \gamma_{23} \end{bmatrix}$$ and deduce from eqn. 2.9

$$G(s) = \boldsymbol{\gamma}.(sI-\mathbf{\Lambda})^{-1}.\boldsymbol{\beta}+D = \boldsymbol{\gamma}\begin{bmatrix} \frac{1}{s+2} & 0 & 0 \\ 0 & \frac{1}{s+3} & 0 \\ 0 & 0 & \frac{1}{s+4} \end{bmatrix}\boldsymbol{\beta}+D$$

Note first that $\boldsymbol{D} = \boldsymbol{0}$. For in eqn. 2.9, as $s \to \infty$, $(s\boldsymbol{I} - \boldsymbol{A})^{-1} = \frac{1}{s}\left(\boldsymbol{I} + \frac{\boldsymbol{A}}{s} + \frac{\boldsymbol{A}^2}{s^2} + \ldots\right) \to \boldsymbol{0}$, so that $\boldsymbol{D} = \underset{s\to\infty}{\mathrm{Lt}}\ \boldsymbol{G}(s)$, and in this example this limit is clearly zero. Next, express the elements of the given $\boldsymbol{G}(s)$ in partial fraction form:

$$g_{11}(s) = \frac{1}{s+2} - \frac{5}{s+3} + \frac{5}{s+4}, \qquad g_{12}(s) = -\frac{1}{s+2} + \frac{4}{s+3} - \frac{3}{s+4},$$

$$g_{21}(s) = -\frac{2}{s+2} + \frac{15}{s+3} - \frac{20}{s+4}, \qquad g_{22}(s) = \frac{2}{s+2} - \frac{12}{s+3} + \frac{12}{s+4}$$

Note that the coefficient of $\frac{1}{s+2}$ in the tentative expression for $\boldsymbol{G}(s)$ is the product (*first* column of $\boldsymbol{\gamma}$).(*first* row of $\boldsymbol{\beta}$) and similarly for $\frac{1}{s+3}, \frac{1}{s+4}$; that these matrix coefficients are therefore necessarily of rank unity; and that since only the product of the column and the row is of importance, the scale of either the row or the column may be fixed arbitrarily. Equating coefficients of the three partial fractions then gives:

$$\begin{bmatrix} \gamma_{11}\beta_{11} & \gamma_{11}\beta_{12} \\ \gamma_{21}\beta_{11} & \gamma_{21}\beta_{12} \end{bmatrix} = \begin{bmatrix} 1 & -1 \\ -2 & 2 \end{bmatrix}; \quad \begin{bmatrix} \gamma_{12}\beta_{21} & \gamma_{12}\beta_{22} \\ \gamma_{22}\beta_{21} & \gamma_{22}\beta_{22} \end{bmatrix} = \begin{bmatrix} -5 & 4 \\ 15 & -12 \end{bmatrix};$$

$$\begin{bmatrix} \gamma_{13}\beta_{31} & \gamma_{13}\beta_{32} \\ \gamma_{23}\beta_{31} & \gamma_{23}\beta_{32} \end{bmatrix} = \begin{bmatrix} 5 & -3 \\ -20 & 12 \end{bmatrix}.$$

Choosing arbitrarily $\gamma_{11} = \gamma_{12} = \gamma_{13} = 1$, we deduce $\beta_{11} = 1, \beta_{12} = -1, \gamma_{21} = -2; \beta_{21} = -5, \beta_{22} = 4, \gamma_{22} = -3; \beta_{31} = 5, \beta_{32} = -3, \gamma_{23} = -4$. Hence a diagonal canonical solution of the problem is given by

$$\boldsymbol{\Lambda} = \begin{bmatrix} -2 & 0 & 0 \\ 0 & -3 & 0 \\ 0 & 0 & -4 \end{bmatrix}; \quad \boldsymbol{\beta} = \begin{bmatrix} 1 & -1 \\ -5 & 4 \\ 5 & -3 \end{bmatrix}; \quad \boldsymbol{\gamma} = \begin{bmatrix} 1 & 1 & 1 \\ -2 & -3 & -4 \end{bmatrix};$$

$$\boldsymbol{D} = 0; \text{ state vector } = \boldsymbol{y}.$$

From this solution, by writing $\boldsymbol{y} = \boldsymbol{E}^{-1}\boldsymbol{x}$ and hence $\boldsymbol{A} = \boldsymbol{E\Lambda E}^{-1}, \boldsymbol{B} = \boldsymbol{E\beta}, \boldsymbol{C} = \boldsymbol{\gamma E}^{-1}$, may be obtained an *infinite number* of state representations of the system transfer function, for $\boldsymbol{E}$ in this transformation is *any* nonsingular square matrix of order 3.

3.4.2 Example 2

A system is represented by the matrices

$$A = \begin{bmatrix} 0 & 1 & 0 & 0 \\ -7 & -8 & -12 & 0 \\ 0 & 0 & 0 & 1 \\ -5 & -5 & -12 & -2 \end{bmatrix}; \quad B = \begin{bmatrix} 6 & 0 \\ -42 & 12 \\ 5 & 0 \\ -35 & 11 \end{bmatrix};$$

$$C = \frac{1}{6}\begin{bmatrix} 1 & 0 & 0 & 0 \\ 0 & 1 & 0 & 0 \end{bmatrix}; \quad D = 0.$$

Show that this representation is not minimal and obtain a minimal representation.

It will be found that rank $[B, AB, A^2B, A^3B] = 3$ and not 4. There is therefore one uncontrollable mode present; in other words one of the elements of y (in the diagonal canonical form) must satisfy an equation of the form $\dot{y}_r = \lambda y_r$ and must moreover, since $y = E^{-1}x$, be some linear combination of the elements of x. Let therefore $y_r = k'.x$, where $k' = [k_1, k_2, k_3, k_4]$. Then $\dot{y}_r = k'.\dot{x} = k'Ax + k'Bu$ and also equals $\lambda y_r = \lambda k'x$. Hence k' and λ must satisfy $k'A = \lambda k'$ and $k'B = 0$.

It is left to the reader to confirm that the six scalar equations involved lead to the unique solution $\lambda = -1$; $k_1 : k_2 : k_3 : k_4 = 17 : 11 : -12 : -12$. The mode exp $(-t)$ is therefore uncontrollable and we may assume $y_r = 17x_1 + 11x_2 - 12x_3 - 12x_4$. Solving this equation for, say, x_4 and substituting for x_4 in the last two state equations (it is absent from the first two) leads to $\dot{x}_3 = \frac{17}{12}x_1 + \frac{11}{12}x_2 - x_3 - \frac{1}{12}y_r + [5 \quad 0]\,u$ in the third equation and $\frac{17}{12}\dot{x}_1 + \frac{11}{12}\dot{x}_2 - \dot{x}_3 - \frac{1}{12}\dot{y}_r = -5x_1 - 5x_2 - 12x_3 - 2(\frac{17}{12}x_1 + \frac{11}{12}x_2 - x_3 - \frac{1}{12}y_r) + [-35 \quad 11]\,u$. On substituting in this last equation for $\dot{x}_1, \dot{x}_2$ and $\dot{x}_3$ and putting $\dot{y}_r = \lambda y_r = -y_r$ it will be found to be an identity and may therefore be omitted. Moreover since a minimal representation excludes noncontrollable modes, we may put $y_r = 0$ whenever it occurs. There remain the three state equations with state vector $x = [x_1, x_2, x_3]'$,

$$\dot{x} = \begin{bmatrix} 0 & 1 & 0 \\ -7 & -8 & -12 \\ \frac{17}{12} & \frac{11}{12} & -1 \end{bmatrix} x + \begin{bmatrix} 6 & 0 \\ -42 & 12 \\ 5 & 0 \end{bmatrix} u$$

The output equation clearly reduces to $v = \frac{1}{6}\begin{bmatrix} 1 & 0 & 0 \\ 0 & 1 & 0 \end{bmatrix} x$ these two equations giving the required minimal representation with $n = 3$.

Having found $\boldsymbol{k}'$ and therefore y_r in terms of $\boldsymbol{x}$, we may use $y_r = \boldsymbol{k}'\boldsymbol{x}$ to substitute for any one of the elements of $\boldsymbol{x}$: x_4 was merely chosen for convenience. Whichever element, x_s, is chosen, the state equation for $\dot{x}_s$ always results in an identity and may therefore be ignored. If there should be more than one uncontrollable mode, there will be a correspondingly multiple solution for λ, and for each value of λ there will be a corresponding value of $\boldsymbol{k}'$: then *each* equation of the form $y_r = \boldsymbol{k}'\boldsymbol{x}$ may be used to eliminate one element of $\boldsymbol{x}$.

The method may be extended to the elimination of *nonobservable* modes by using the duality principle: any nonobservable mode of the system $(\boldsymbol{A}, \boldsymbol{B}, \boldsymbol{C}, \boldsymbol{D})$ is a noncontrollable mode of the system $(\boldsymbol{A}', \boldsymbol{C}', \boldsymbol{B}', \boldsymbol{D})$.

An alternative method of solving this problem is of course to convert the transformation to diagonal canonical form and to omit the mode corresponding to a null row of $\boldsymbol{\beta}$. But to do this requires finding *all* the eigenvalues of $\boldsymbol{A}$, as well as their associated eigenvectors, and inverting $\boldsymbol{E}$: the suggested method reduces this work to a minimum by merely finding the eigenvalues associated with noncontrollable modes and the associated $\boldsymbol{k}'$, which, since $\boldsymbol{y} = \boldsymbol{E}^{-1}\boldsymbol{x}$, are the associated rows of $\boldsymbol{E}^{-1}$.

It is left to the reader to verify that, in Example 2, both the original fourth-order representation and the minimal third-order representation lead to the same value of transfer matrix and that this value is precisely the value of $G(s)$ in Example 1.

3.5 Examples 3

1. If $\boldsymbol{A} = \begin{bmatrix} -3 & 1 \\ -2 & 1{\cdot}5 \end{bmatrix}$ and $\boldsymbol{B} = \begin{bmatrix} 1 \\ 4 \end{bmatrix}$ show that the system is not completely controllable. If, in addition, $\boldsymbol{C} = [c_1, c_2]$, find the condition that the system shall be observable. Confirm your findings in each case by using criteria in the domain of complex frequency.

Converting the state equation to the diagonal canonical form, find the two natural modes and identify the noncontrollable mode.

2. Return to the linearised form of the equations for Example 1.3 and investigate the controllability of the system if (i) u_1 is absent, (ii) u_2 is absent.

3. In Example 2 of Section 3.4.2, suppose that $\boldsymbol{C} = [d, e, f]$ and that $\boldsymbol{A}$ and $\boldsymbol{B}$ are the reduced (third-order state vector) matrices given. Find the relation or relations that must hold between d, e, f if the system is to be *non*observable. Select any numerical values of d, e, f such that one

and only one of the three state modes shall be *non*observable. With these values, obtain a second-order representation of the system and show that the transfer matrix of this new representation is identical with that for the original third- (or fourth-) order representation.

4. In a system with two input and two output quantities, the transfer matrix is given by $\boldsymbol{T}(s)$, where $\boldsymbol{V}(s) = \boldsymbol{T}(s)\,\boldsymbol{U}(s)$ and

$$\boldsymbol{T}(s) = \frac{1}{(s+1)(s+2)(s+3)}\begin{bmatrix}(3s+7)(s+3) & 2s+5\\(s+1)(s+3) & s+1\end{bmatrix}$$

Show that if the method of the Appendix to Chapter 1 is applied to the separate equations for $V_1(s)$ and $V_2(s)$, the system is representable by a sixth-order state equation. Using the technique of Section 3.4.2 or otherwise (e.g. by diagonal canonical form transformation) reduce the order of this transformation until a minimum-order representation (of order 3) is obtained and confirm that the transfer matrix for this representation is indeed $\boldsymbol{T}(s)$.

5. Show that the transfer matrix $\boldsymbol{T}(s)$ of Example 4 can be realised by a forward path of transfer matrix

$$\boldsymbol{L}(s) = \frac{1}{(s+k)(s+8-k)}\begin{bmatrix}3s-8+3k & 2\\s-4+k & 1\end{bmatrix}$$

bridged by a negative feed-back path of transfer matrix

$$\boldsymbol{M} = \begin{bmatrix}k-7 & 11-2k\\-(k-7)(k-5) & (3k-17)(k-5)\end{bmatrix}$$

whatever the value of k. If the equations for this feedback network are written as $(\boldsymbol{I}+\boldsymbol{LM})\,\boldsymbol{V} = \boldsymbol{LU}$, show that a fourth-order representation results. If the same equation is modified to $(\boldsymbol{L}^{-1}+\boldsymbol{M})\,\boldsymbol{V} = \boldsymbol{U}$, show that a (minimal) third-order representation results; if the latter is not identical to that found in Example 4, show that it can be obtained from it by a nonsingular transformation of the state vector.

Chapter 4

On stability

4.1 Introduction

The system to be analysed is assumed to have the state equation, eqn. 1.2:

$$\dot{\boldsymbol{x}} = \boldsymbol{f}(\boldsymbol{x}, \boldsymbol{u}, t) \tag{4.1}$$

corresponding to a possibly nonlinear, time-dependent system, though of course linear and/or autonomous systems may be considered as particular cases.

In the majority of cases the stability, to be defined later, of a linear system is not affected by the input. On the other hand the stability of nonlinear systems is usually affected by the input. It is therefore desirable to include the input in discussing the time behaviour of the system. There is unfortunately no general theory of system stability in the presence of arbitrary inputs: we shall therefore compromise and assume that if the input is not zero, then it is constant.

In the majority of cases a linear system has only one point of equilibrium, though there are exceptions. A nonlinear system on the other hand may have a number of equilibrium points and stability has to be investigated in the neighbourhood of each.

4.1.1 Equilibrium

An equilibrium point in the state-space is a point where $\dot{\boldsymbol{x}} = \boldsymbol{0}$. More precisely: if, for any constant input vector $\boldsymbol{u}(t) = \boldsymbol{U}$ there exists a point in the state-space $\boldsymbol{x}(t) = \boldsymbol{x}_e$ = constant, such that at this point $\dot{\boldsymbol{x}}(t) = \boldsymbol{0}$ for all t, then this point is an equilibrium point of the system corresponding to the particular input $\boldsymbol{U}$.

Applying this definition first to a linear system, with state equation

$$\dot{x} = Ax + Bu \tag{4.2}$$

we require

$$Ax_e + BU = 0 \tag{4.3}$$

and, provided only that A is nonsingular, this has the *unique* solution

$$x_e = A^{-1}BU$$

Note however that in a time-dependent system it is probable that $A^{-1}B$ will be time-dependent: in this case the only solution *valid for all t* will be $x_e = 0 = U$, namely the origin of the state-space with zero input.

On the other hand, if A is singular, let rank $A = R < n$. (We are *not* implying tha' the *n state equations* are of rank R: they are still assumed to be of rank n). Suppose for simplicity that the first R rows of A are l.i.; then any later row, say the rth, is a linear combination of the first R. Hence, subtracting this linear combination of the first R *state equations* from the rth equation, there will result some linear combination of the elements of U equated to zero. Taking $r = (R + 1)$, $(R + 2), \ldots, n$, will result therefore in $(n - R)$ such linear, homogeneous equations in the elements of U which must be satisfied if x_e is to be found. (Note that if the rank of these equations is greater than or equal to m, the only solution is $U = 0$.) Provided that U satisfies these constraints, then the first R equations of eqn. 4.3 may be solved for so some R of the n elements of x_e, *allocating completely arbitrary values to the other $(n - R)$ elements*. This shows that x_e is not a discrete point but a continuum of points, x_e lying on a 'line', 'plane', . . . , in the n-dimensional state space according as $(n - R) = 1, 2, \ldots$. In brief, x_e has $(n - R)$ degrees of freedom.

This Chapter will deal almost entirely with systems for which the equilibrium points are discrete points.

For a nonlinear system represented by eqn. 4.1, any equilibrium point must satisfy

$$f(x_e, U, t) = 0 \qquad \text{for all } t \tag{4.4}$$

The number of solutions depends entirely upon the nature of the function elements of f and no general statement is possible.

In studying the stability of a system in relation to one of its equilibrium points, it is convenient to shift the origin of the state space to this point. Thus, writing

$$z = x - x_e \qquad \text{and therefore} \qquad \dot{z} = \dot{x}$$

the state equations, referred to $x = x_e$ as origin, become

$$\dot{z} = f(z + x_e, U, t) \equiv g(z, U, t) \quad \text{say} \tag{4.5}$$

and it is obvious that $z = \boldsymbol{0}$ is an equilibrium point since $\boldsymbol{x}_e$ satisfies eqn. 4.4.

Similarly, for a linear system, we deduce

$$\dot{z} = \boldsymbol{A}.(z + \boldsymbol{x}_e) + \boldsymbol{BU}$$

or

$$\dot{z} = \boldsymbol{A}z \qquad (4.6)$$

since $\boldsymbol{x}_e$ satisfies eqn. 4.3.

4.2 Concepts of stability

If the state of a system is $\boldsymbol{x} = \boldsymbol{x}_e$, then the system stays in this state indefinitely since $\dot{\boldsymbol{x}} = \boldsymbol{0}$ for all t. If, however, the system is disturbed, however minutely, from this position, the state vector $\boldsymbol{x}$ will in general vary with time, a variation which it is convenient to visualise as the point point $\boldsymbol{x}(t)$ describing a trajectory in the state space. The nature of this trajectory, i.e., the nature of the subsequent behaviour of the system when disturbed from its equilibrium state, determines the stability or instability of the equilibrium point in question. This subsequent behaviour may depend upon the direction and/or magnitude of the initial displacement from the equilibrium point, but, when this is given, is entirely controlled by the dynamic equations of the system, namely the state equation. It will be assumed throughout the analysis that, as outlined in the previous section, the equilibrium point considered is taken a as the origin of the state space.

4.2.1 Some definitions

Various authors have suggested over thirty definitions of various types of stability! We shall only deal with a few of the more important ones.

Suppose the system is represented by eqn. 4.5 or, if linear, by eqn. 4.6. Suppose that at $t = t_0$ the system is displaced to $z = z_0$ and that subsequently, for $t \geqslant t_0$, the trajectory, which depends in general on t_0, z_0, U and t, is given by

$$z(t) = \boldsymbol{h}(z_0, U, t_0, t)$$

where $\boldsymbol{h}$ is some function-vector of time t involving the constants z_0, U, t_0 as parameters.

Definition (i) The equilibrium point (at the origin) is said to be stable if for *every* positive number α there exists another positive number $\beta(\alpha, t_0)$ such that the inequality norm $\boldsymbol{z}_0 < \beta$ implies the inequality norm $\boldsymbol{h} < \alpha$.

The (Euclidean) norm of a vector is the positive square root of the sum of the squares of its elements and corresponds to distance from the origin in 3-dimensional space. Thus the equilibrium point is stable if for *every* 'distance' α from the origin there exists a 'distance' β (clearly $\beta \leqslant \alpha$) such that, if the state-vector starts at a 'distance' $< \beta$, it will at no subsequent time lie at a greater 'distance' than α.

For stability the definition must apply to *every* α, however small. If α is small enough, the behaviour of the system may be approximated to by its linearised form eqn. 4.6. If the system is autonomous and $\boldsymbol{z} = \boldsymbol{z}_0$ at $t = t_0 = 0$, then the solution of this equation, in Laplacian form, is

$$Z(s) = \text{adj}(s\boldsymbol{I} - \boldsymbol{A}).\boldsymbol{z}_0/\det(s\boldsymbol{I} - \boldsymbol{A})$$

and partial fraction analysis followed by inverse Laplace transformation shows that, provided the eigenvalues of $\boldsymbol{A}$ have negative real parts, every element of $\boldsymbol{z}(t)$ will remain bounded and ultimately tend to zero. Thus by making $\boldsymbol{z}_0$ small enough in norm we can make norm $\boldsymbol{z}(t)$, for any positive t, as small as we please. The system is then said to be *stable in the small*. This will still be true if $\boldsymbol{A}$ has one or more *nonrepeated* eigenvalues on the imaginary axis. The actual system may however be stable in the small without being stable, since the linearised form is only an approximation to the true system.

Definition (ii) The equilibrium point is said to be *asymptotically stable* if (*a*) it is stable, (*b*) the trajectory returns to the origin at some subsequent time $t = T > t_0$. (T may be finite or infinite). It follows that the linearised form of an autonomous system is asymptotically stable if and only if all the eigenvalues of $\boldsymbol{A}$ have negative real parts.

Definition (iii) If the equilibrium point is asymptotically stable, the assembly of points $\boldsymbol{z}_0$ such that a trajectory originating at $\boldsymbol{z} = \boldsymbol{z}_0$ ultimately reaches the origin is called the *domain of attraction* of the equilibrium point. If this domain of attraction includes every point in the state space, the equilibrium point is said to be *globally asymptotically stable*.

Definition (iv) If the equilibrium point is not stable according to definition (i), it is said to be unstable.

4.2.1.1 Comments

If a trajectory at any time reaches the origin, then it terminates there, since at the origin $\dot{z} = 0$.

It is clearly desirable that the equilibrium point of a control system should be asymptotically stable rather than merely stable; otherwise even a small disturbance may result in a relatively large sustained oscillation of the state vector. It is, moreover, desirable that its domain of attraction should be at least large enough to accommodate all states likely to occur in normal operation. Ideally the asymptotic stability should be global, but this implies the absence of any other equilibrium points, stable or not, and this is usually difficult to achieve unless the system is assumed linear throughout the state space.

4.3 Criteria of stability

The criteria of stability for linear systems associated with Routh, Hurwitz, Nyquist and Bode do not require the *solution* of the system equations. With nonlinear systems, the difficulty or impossibility of solving these equations analytically demands, even more forcibly, stability criteria avoiding the process of solution. As might be expected, the search for such criteria is an arduous process.

4.3.1 First method of Liapunov

Among other results, Liapunov proved rigorously that if nonlinear equations are linearised in the neighbourhood of an equilibrium point to yield equations of the form

$$\dot{z} = Az + 0_2(z)$$

where $0_2(z)$ denotes terms of the second or higher degree in the elements of z, then provided that no eigenvalues of A have zero real part, the nature of the equilibrium of the nonlinear system is the same as that of the linear representation $\dot{z} = Az$. This we have already assumed. More interesting is:

Theorem 4.1 If any of the eigenvalues of A have a zero real part, the others having negative real parts, the nature of the equilibrium of the origin is determined by the eigenvalues of the Jacobian matrix (from which A is obtained) evaluated *in the neighbourhood of*, instead of at, the origin. The following very simple example will serve to illustrate a number of points in connection with the above definitions as well as to give an application of this theorem.

4.3.2 Example

Consider a simple pendulum consisting of a light rod of negligible mass and length h, freely pivoted at one end and carrying a point mass m at the other. The pendulum oscillates in a vertical plane and the deflection of the rod from the downward vertical is $\theta(t)$. In the absence of any forces other than gravity, the equation of motion is

$$mh\ddot{\theta} = -mg \sin \theta$$

or, taking for simplicity $h = g$, we obtain $\ddot{\theta} = -\sin \theta$. Writing this in state vector form with $x_1 = \theta, x_2 = \dot{\theta} = \dot{x}_1$, there results $\dot{x} = \begin{bmatrix} x_2 \\ -\sin x_1 \end{bmatrix}$. The system equilibrium points are clearly given by $x_2 = 0, x_1 = N\pi$, where N is any integer. Note that these points are infinite in number in the state space (the plane $x_1 0 x_2$) although they correspond to only two rest positions of the pendulum depending on whether N is even or odd.

Shifting the origin to $x_1 = N\pi, x_2 = 0$, by writing $z_1 = x_1 - N\pi$, $z_2 = x_2$, gives

$$\dot{z} = \begin{bmatrix} z_2 \\ -\sin(z_1 + N\pi) \end{bmatrix} = \begin{bmatrix} z_2 \\ (-1)^{N+1} \sin z_1 \end{bmatrix} \equiv f(z)$$

The Jacobian matrix $J(f: z)$ is therefore obtained as

$$J = \begin{bmatrix} 0 & 1 \\ (-1)^{N+1} \cos z_1 & 0 \end{bmatrix} \quad \text{or, with } z_1 = 0, A = \begin{bmatrix} 0 & 1 \\ (-1)^{N+1} & 0 \end{bmatrix}$$

If N is odd (upper vertical position of rod) the eigenvalues of A are ± 1 and these equilibrium points are unstable since one eigenvalue is positive. If N is even (lower vertical position of rod) the eigenvalues of A are $\pm j$ and the *linearised representation* is stable, but not necessarily the actual nonlinear system. This uncertainty is, moreover, unresolved by using Liapunov's theorem: making z_1 small instead of zero still gives imaginary eigenvalues.

Consider in more detail the case $N = 0$, say. Integrating the original equation of motion after multiplying it by $2\dot{\theta}$ gives

$$\dot{\theta}^2 - 2\cos\theta = \text{constant}$$

which is both a statement of the conservation of energy principle and the equation of all possible trajectories. It may be written as

$$x_2^2 + 4\sin^2\frac{x_1}{2} = \text{constant} = x_{20}^2 + 4\sin^2\frac{x_{10}}{2} \qquad (4.7)$$

By making x_{10} and x_{20} small enough, i.e. by starting near enough to $\theta = 0$ with a low enough velocity, the right hand side may be made as small as we please. It then follows that x_1 and x_2 are also very small for all t and the trajectories simplify to $x_2^2 + x_1^2 = \text{constant}$, namely small circles with the origin (the equilibrium point) as centre. As norm $\boldsymbol{x}_0$ is increased, the trajectory at first becomes a deformed circle (curves a, b in Fig. 4.1) but if the constant in eqn. 4.7 equals 4, this equation simplifies to $x_2 = \pm 2\cos\dfrac{x_1}{2}$ (curve c): the pendulum *just* reaches one or other of the neighbouring equilibrium points $x_1 = \pm\pi, x_2 = 0$ and stays there. If the constant is increased still more, the trajectory (curves d) is such that x_2 never vanishes and the pendulum goes on rotating indefinitely, clockwise or anticlockwise according to the initial conditions.

All equilibrium points $x_1 = N\pi, x_2 = 0$, with N even, are clearly stable by definition (i). For however small or however large we choose

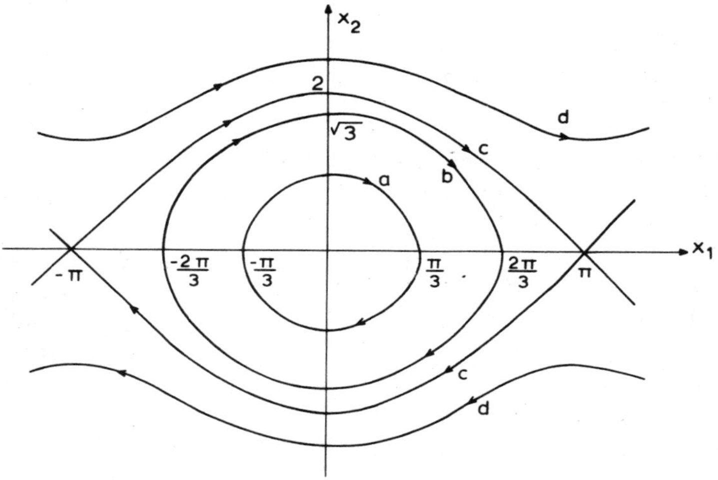

Fig. 4.1 Typical state-plane trajectories for simple pendulum problem. Arrows indicate direction of increasing time

α, there is always a value of β (<2) such that if we start within a circle of radius β, the trajectory remains within or on radius β and therefore within radius α. To distinguish between initial conditions lying inside the loops c in Fig. 4.1 and those lying above or below these loops, we qualify this stability by saying that these equilibrium points are *locally stable* for the former (since then the trajectory is bounded by one of the loops) but *unstable in the large* for the latter, (since the trajectories now extend to infinity).

No equilibrium point, whether N be even or odd, is asymptotically stable: for no trajectory ever goes through an N-even equilibrium point, and although trajectory c brings the state to one or other of the N-odd equilibrium points, these are unstable and therefore cannot be asymptotically stable.

Finally, if we modify the problem by introducing a resistive force due to the medium in which the pendulum swings, say a force $k\dot{\theta}\,|\dot{\theta}\,|\,.\,mg$, the equation of motion becomes

$$\ddot{\theta} = -\sin\theta - k\dot{\theta}\,|\dot{\theta}\,| \qquad (k>0)$$

(The form $\dot{\theta}\,|\dot{\theta}\,|$ and not $\dot{\theta}^2$ is used since the force changes sign with $\dot{\theta}$.) We deduce

$$\dot{\boldsymbol{x}} = \begin{bmatrix} x_2 \\ -(\sin x_1 + kx_2\,|x_2\,|) \end{bmatrix}$$

and clearly the equilibrium points are unchanged. The Jacobian matrix becomes

$$J(\dot{\boldsymbol{x}}:\boldsymbol{x}) = \begin{bmatrix} 0 & 1 \\ -\cos x_1 & -2k\,|x_2\,| \end{bmatrix}$$

which, evaluated at the origin of $\boldsymbol{x}$ (i.e. $N=0$) gives as before a matrix $\boldsymbol{A}$ with eigenvalues $\pm j$. But if $\boldsymbol{J}$ is evaluated *near* the origin (using the Liapunov theorem) we find, to a first order of small quantities (x_1 and x_2), that $\boldsymbol{A} = \begin{bmatrix} 0 & 1 \\ -1 & -2k\,|x_2\,| \end{bmatrix}$ of which the eigenvalues, given by $\lambda^2 + 2k\,|x_2\,|\lambda + 1 = 0$, clearly have negative real parts, so that, using the theorem, the origin is a stable equilibrium point for the system. (If k could be made negative, the origin would be an unstable equilibrium point.) Since the eigenvalues in the neighbourhood of the equilibrium point all have negative real parts, the equilibrium point is now not merely stable but asymptotically stable: the pendulum will ultimately come to rest at $\theta = x_1 = 0, \dot{\theta} = x_2 = 0$, provided that the initial conditions are not such as to make it swing through $\theta = \pm\,\pi$, for in this case it will ultimately settle down at *another* asymptotically stable point such as $x_1 = \pm 2\pi, \pm 4\pi, \ldots; x_2 = 0$. There is therefore a finite *domain of attraction* [Definition (iii)].

4.3.2 The second method of Liapunov: Liapunov functions

The reader will probably be familiar with the concept that if a system has an asymptotically-stable equilibrium point, then if it starts within the domain of attraction, the stored energy of the system will subsequently decrease with time, though not necessarily monotonically, until it reaches a minimum at the equilibrium point. Thus the damped simple pendulum just considered will have a gravitational potential energy which is a minimum at $\theta = 0$, but only if the damping is high enough will the approach to this position be monotonic. A Liapunov function is similar in nature to an energy function but *always* decreases monotonically with respect to time. Some of the essential theorems relating to such functions will first be given, followed by a discussion of their advantages and disadvantages. The origin of $\boldsymbol{x}$ is assumed to be the equilibrium point considered.

4.3.2.1 Theorems

Theorem 4.2 If there exists a scalar function $V(\boldsymbol{x}, t)$–a Liapunov function–with continuous first partial derivatives and satisfying:

(*a*) $V(\boldsymbol{0}, t) = 0$ for all t

(*b*) $V(\boldsymbol{x}, t) > 0$ for all $\boldsymbol{x} \neq \boldsymbol{0}$ and for all t

(*c*) $\dot{V}(\boldsymbol{0}, t) = 0$ for all t

(*d*) $\dot{V}(\boldsymbol{x}, t) < 0$ for all $\boldsymbol{x} \neq \boldsymbol{0}$ and for all t

(*e*) $V(\boldsymbol{x}, t) \rightarrow \infty$ as norm $\boldsymbol{x} \rightarrow \infty$, for all t

then the origin is globally, asymptotically stable. (In autonomous systems V is not explicitly a function of t and the phrase 'for all t' is then superfluous.)

Theorem 4.3 If in Theorem 4.2 condition (*d*) is slackened to $\dot{V}(x, t) \leqslant 0$ then the origin is stable but not necessarily asymptotically stable. Corollary: If conditions (*a*) to (*d*) hold within a domain D surrounding the origin, then the origin is asymptotically stable in Theorem 4.2, stable in Theorem 4.3, but not necessarily globally so.

Theorem 4.4 If, in Theorem 4.2, condition (*d*) is reversed to $\dot{V}(x, t) > 0$

and if, with this modification, (*a*) to (*d*) are valid within some domain D surrounding the origin, then the origin is unstable.

4.3.2.2 Comments

(i) The value of $\dot{V}$ in the above theorems is to be obtained from

$$\dot{V} = \frac{\partial V}{\partial t} + \sum_{r=1}^{n} \frac{\partial V}{\partial x_r} \dot{x}_r$$

the values of the $\dot{x}_r$ being obtained from $\dot{\boldsymbol{x}} = \boldsymbol{g}(\boldsymbol{x}, U, t)$, see eqn. 4.5.

(ii) A Liapunov function, unlike a system energy function, is not unique for a given system and a given equilibrium point. If no function can be found to justify the use of any of the above theorems, this pooves nothing, except perhaps that the investigator has not been thorough enough in his search!

(iii) Conditions (*a*) and (*b*) may be telescoped into the single phrase 'V is positive-definite' and (*c*) and (*d*) into '$\dot{V}$ is negative-definite' (Theorem 4.2), '$\dot{V}$ is negative semi-definite' (Theorem 4.3), '$\dot{V}$ is positive-definite' (Theorem 4.4).

(iv) If we consider V as a sum of terms of degree 0, 1, 2 . . . in the elements of $\boldsymbol{x}$, then, by condition (*a*), the terms of degree 0 must be absent. The terms of degree 1 must also be absent, for near the origin they dominate the expression for V, but, being linear, they change sign with $\boldsymbol{x}$, so that, if they are present, V cannot be positive-definite near the origin. Thus the lowest permissible degree is 2, and these quadratic terms, since they determine V near the origin, must be positive-definite. If the quadratic terms are absent then the cubic terms are also absent, for the same reason as the linear terms. Thus the terms of lowest degree must be of degree 2, 4, . . .

(v) It is important to distinguish between a domain of attraction and a domain within which, say, V is positive-definite and $\dot{V}$ negative-definite. For it is quite possible for a trajectory to originate within the latter but, in the process of time, to go outside this domain where $\dot{V}$ may be positive so that V may $\to \infty$ as $t \to \infty$. Such a starting point clearly does not lie in the domain of attraction of the origin. Conversely, a trajectory may start in a region where, say, both V and $\dot{V}$ are positive, but cross over into a region where $\dot{V}$ is negative and ultimately terminate at the origin. Thus it would appear that Liapunov functions, as far as the signs of V and $\dot{V}$ are concerned, do not determine domains of attraction, unless of course the origin is globally, asymptotically stable.

(vi) The weakness of the method lies, of course, in the fact that there is normally no systematic method of finding a Liapunov function for a particular nonlinear system. One must try various forms, possibly starting with a quadratic form and adding higher terms if found useful, spurred on by the fact that if the system is stable, a Liapunov function exists, satisfying Theorems 4.2 or 4.3. Various authors have made helpful suggestions in connection with particular types of nonlinearity, others have suggested methods that may be useful in general: we shall briefly present the more useful of these. But the hard fact remains that in any but the simplest systems the search for a Liapunov function remains arduous: our only consolation lies in the fact that the only other way of assessing the stability of an equilibrium point is by a numerical solution of the nonlinear equations, using a digital or analogue computer and repeating the process for a sufficient variety of starting conditions to cover likely practical situations. Note that if $n = 6$, say, and the probable range of each x_r at $t = t_0$ is broken up into ten intervals, we have to solve six nonlinear first order, nonlinear equations nearly two million times! Perhaps a little time spent searching for a Liapunov function is well spent!

4.3.3 Isoclinals: A graphical method for second order systems

If we are dealing with a second-order, autonomous system ($n = 2$), the sketching of trajectories in the state-space (i.e. state-plane) may usually be carried out fairly simply by first plotting in this plane a number of *isoclinals* or curves of constant trajectory slope. Taking the state equations as $\dot{x}_1 = f_1(x_1, x_2), \dot{x}_2 = f_2(x_1, x_2)$, we obtain, on dividing the second by the first

$$\frac{dx_2}{dx_1} = \frac{f_2(x_1, x_2)}{f_1(x_1, x_2)} \equiv S, \text{ say}$$

If then, for a particular value of S, we plot the curve $f_2 = Sf_1$, any trajectory must cut this curve with slope S. If a number of these isoclinals are plotted, for different values of S, then, with any starting point, it is not difficult to sketch the subsequent trajectory: the more isoclinals are plotted, the greater the accuracy of the trajectory sketch.

Note that at any equilibrium point both f_1 and f_2 vanish, so that S is indeterminate: any number of isoclinals may therefore intersect at an equilibrium point. Note also that time t is absent explicitly from the equations: at any point of the state-plane we may however determine the sign of $\dot{x}_1 = f_1$ and $\dot{x}_2 = f_2$ and thus determine the direction of travel along the trajectory as time increases.

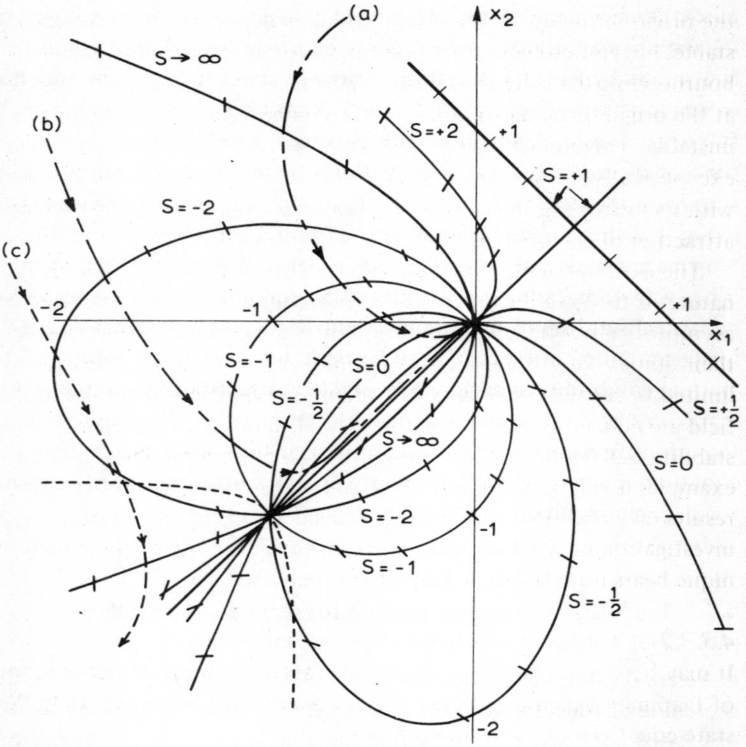

Fig. 4.2 Isoclinals for $\dot{x}_1 = -x_1 - x_2^2, \dot{x}_2 = -x_2 - x_1^2$

4.3.3.1 Example

Suppose $\dot{x}_1 = -x_1 - x_2^2, \dot{x}_2 = -x_2 - x_1^2$. The equilibrium points are given by $x_1 = -x_2^2 = -x_1^4$. Hence either $x_1 = x_2 = 0$ or $x_1 = x_2 = -1$. The isoclinals are given by $x_1^2 + x_2 = S(x_2^2 + x_1)$ and are therefore conic sections through both equilibrium points. The isoclinals for $S = 0, \pm\frac{1}{2}, \pm 1, \pm 2, \infty$, are shown with reasonable accuracy in Fig. 4.2. For ease of trajectory plotting, on each isoclinal are drawn short strokes of slope S.

To every trajectory possible above the line $x_2 = x_1$ there is clearly an image trajectory in this line. Three trajectories are shown, (*a*), (*b*) and (*c*). All three, if extrapolated backwards in time (against the arrows) tend to infinity with unity slope. It is clear that (*a*) and (*b*), and indeed any trajectory initiated in the region between (*b*) and its mirror image in $x_2 = x_1$, will ultimately terminate in the origin. Trajectory (*c*), on the other hand, clearly travels to infinity. All trajectories initiated near the origin terminate at the origin, which is therefore asymptotically stable, but not globally. Nearly all trajectories initiated in the neighbourhood of the other equilibrium point $(-1, -1)$ either terminate at the origin or travel to infinity: this equilibrium point is therefore unstable. There must be a critical trajectory, between (*b*) and (*c*) which, exceptionally, terminates at $(-1, -1)$: this critical trajectory, together with its mirror image in $x_2 = x_1$, forms the boundary of the domain of attraction of the origin.

The isoclinal method is therefore capable of determining both the nature of the equilibrium at any equilibrium point and, in the case of asymptotically stable equilibrium points, with reasonable accuracy, their domain of attraction. The method, unfortunately, is clearly limited to second-order ($n = 2$) systems; nevertheless examples in this field are educative in so far as they reveal the various patterns of stability behaviour of which nonlinear systems are capable. For this example, it is left as a simple exercise for the reader to confirm the results on stability 'in the small' obtained above, by linearisation and investigation of the Jacobian matrix eigenvalues. The isoclinal technique bears no relation to Liapunov functions.

4.3.3.2 A Liapunov function for this example

It may be of interest to try, in this example, the simplest possible form of Liapunov function: $V = x_1^2 + x_2^2$. On differentiating and using the state equations, there results

$$\dot{V} = 2x_1\dot{x}_1 + 2x_2\dot{x}_2 = -2(x_1^2 + x_2^2 + x_1x_2^2 + x_2x_1^2)$$

V is positive definite throughout the state space. $\dot{V}$ is negative in the domain

$$h(x_1, x_2) \equiv x_1x_2^2 + x_2x_1^2 + x_1^2 + x_2^2 > 0$$

or, in polar form, writing $x_1 = r\cos\theta, x_2 = r\sin\theta$,

$$h = r^2[1 + r\sin\theta\cos\theta\,(\sin\theta + \cos\theta)] > 0$$

This domain is found to extend over the state-plane with the exception of (*a*) the origin, as an isolated point, where $h = 0$ but is positive near the origin; (*b*) three zones extending to infinity, one in the range $\pi < \theta < 3\pi/2$ and two others, mirror images in $x_2 = x_1$, in the ranges $\pi/2 < \theta < 3\pi/4$ and $-\pi/4 < \theta < 0$. Only the boundary of the first zone is shown in Fig. 4.2, as a dotted line, since the other two lie too far from the origin. By the corollary to Theorem 4.3, the origin is asymptotically stable. No conclusion can be reached on the basis of this tentative V-function about the nature of the stability at $(-1, -1)$ since $\dot{V}$ is positive on one side of this point but negative on the other.

Note that a trajectory through say, $(-2, 0)$, where $V > 0$ and $\dot{V} < 0$, will soon cross the dotted-line boundary $h = 0$, and migrate into a region where $\dot{V}$ is positive, ultimately going to infinity. This instance reinforces earlier remarks stressing the difference between a region where $V > 0$ and $\dot{V} < 0$ and a domain of attraction: the point $(-2, 0)$ lies in the former but clearly not in the latter.

It is of course probable that a more elaborate form of the V function than the very simple form suggested may yield more conclusive information, for instance regarding the stability or otherwise of the point $(-1, -1)$. Indeed it might be hoped that there exists a V function such that $V > 0$ over the whole state plane but such that $\dot{V}$ is negative within the whole domain of attraction of the origin but positive at every point outside this domain. The problem is to find such a function!

4.4 Aids to finding Liapunov functions

4.4.1 A linear system theorem

Although the stability of linear representations has already been dealt with by investigating the real parts of the eigenvalues of the matrix $\boldsymbol{A}$, the following theorem may be useful in so far as it does not require the finding of these eigenvalues.

Theorem 4.5 A necessary and sufficient condition for the origin to be a globally asymptotically stable equilibrium point is that there should exist a positive definite symmetric matrix $\boldsymbol{P}$ satisfying $\boldsymbol{A}'\boldsymbol{P} + \boldsymbol{P}\boldsymbol{A} = -\boldsymbol{I}$; if this is the case, $\boldsymbol{x}'\boldsymbol{P}\boldsymbol{x}$ is a Liapunov function for the representation.

Note that since $\boldsymbol{P}$ is symmetric, it contains $n(n+1)/2$ arbitrary elements. Since $(\boldsymbol{A}'\boldsymbol{P} + \boldsymbol{P}\boldsymbol{A})$ is also symmetric, equating it to $-\boldsymbol{I}$ gives $n(n+1)/2$ linear equations for these elements. It will be found that the solution of these equations only yields a positive definite matrix $\boldsymbol{P}$ if the eigenvalues of $\boldsymbol{A}$ all have negative real parts. $\boldsymbol{A}$ is assumed autonomous.

4.4.2 Krasovskii's theorem

Let the state equation of an autonomous but nonlinear system be written, adapting eqn. 4.5, as $\dot{\boldsymbol{z}} = \boldsymbol{g}(\boldsymbol{z}, \boldsymbol{U})$, where $\boldsymbol{z} = \boldsymbol{0}$ is an equilibrium point. Form the Jacobian matrix $J(g:z)$, such that $j_{rs} = \dfrac{\partial g_r}{\partial z_s}$. Form the symmetric matrix $K = J + J'$. Then if K is negative definite the equilibrium point $z = 0$ is asymptotically stable and a Liapunov function for the system is $V(z) = g'g$. If in addition $g'g \to \infty$ as norm $z \to \infty$, the origin is asymptotically stable in the large.

4.4.3 The Shultz-Gibson variable gradient method

Let the state equation referred to an equilibrium point as origin be given by eqn. 4.5: $\dot{z} = g(z, U, t)$. Assume a Liapunov function $V(z)$ which is a function of z but not of t. Define the gradient of V as the row matrix

$$\text{grad } V = \left[\frac{\partial V}{\partial z_1}, \frac{\partial V}{\partial z_2}, \ldots, \frac{\partial V}{\partial z_n}\right]$$

Then

$$\dot{V} = (\text{grad } V).\dot{z} = (\text{grad } V).g$$

Moreover, since V vanishes at the origin (a necessary property of a Liapunov function) its value at any other point in the state space may be derived from grad V by line integration from the origin to this other point:

$$V(z) = \int_0^Z (\text{grad } V).dz$$

the integral being evaluated along some path in the state space joining the origin to some point $z = Z$. It may be shown that the value of the integral is independent of the path chosen provided

$$\frac{\partial(\text{grad } V)_r}{\partial z_s} = \frac{\partial(\text{grad } V)_s}{\partial z_r}$$

for all possible r, s, from 1 to n, a constraint upon the form of grad V selected which we assume to be satisfied. The simplest path of integration then consists of integrating partially with respect to z_1 from the origin to $(Z_1, 0, 0, \ldots, 0)$; then with respect to z_2 from $(Z_1, 0, 0, \ldots, 0)$ to $(Z_1, Z_2, 0, \ldots, 0)$ etc., concluding with an integration with respect to z_n from $(Z_1, Z_2, Z_3, \ldots, Z_{n-1}, 0)$ to $(Z_1, Z_2, Z_3, \ldots, Z_{n-1}, Z_n)$.

Hence, starting from some arbitrary (grad V) vector, we may, as shown above, obtain $\dot{V}$ by differentiation and using the state equation, and obtain V by line integration. (Grad V) must be chosen to obey the constraint given above, which is an n-dimensional extension of curl(grad V) = 0 in 3-dimensional vector theory. Moreover, since V is at least quadratic in the elements of z, (grad V) is at least linear in these elements. Finally, having obtained V and $\dot{V}$, the coefficients in (grad V) which appear in these two expressions must be adjusted to make $V > 0$ and $\dot{V} < 0$ over as wide a domain as possible.

Examples illustrating the use of this method are either so simple as to render the method unnecessary (the same results being obtainable by assuming a low degree V function in the first place) or too lengthy for the present text. For proofs of the various theorems in this Chapter and for a more thorough presentation of the gradient method, we refer the reader to Ogata and Gibson.

4.5 References

1 GIBSON, J.E.: 'Nonlinear automatic control' (McGraw Hill, 1963)
2 OGATA, K.: 'State space analysis of control systems' (Prentice-Hall, 1967)

4.6 Examples 4

1. Consider the normalised form of Van der Pol's equation

$$\ddot{x} + (x^2 - 1)\dot{x} + x = 0$$

which illustrates the behaviour of some simple types of oscillator.

Put this equation into state-vector form with $x_1 = x, x_2 = \dot{x}$. Show that the only point of equilibrium in the state plane is the origin and that this equilibrium point is unstable.

Sketch isoclinals over the range $-4 \leqslant x_1 \leqslant 4, -4 \leqslant x_2 \leqslant 4$, using values of $S = 0, \pm\frac{1}{2}, \pm 1, \pm 2, \infty$ (or further values if judged necessary). Deduce from the diagram that, wherever the starting point may be, the trajectory always tends to a finite limit-cycle. (Does the form of the equation suggest this?)

2. Consider Duffing's nonlinear equation

$$\ddot{x} + a\dot{x} + bx + cx^3 = 0 \qquad (a, b > 0)$$

Put it into state vector form and investigate the nature of the equilibrium at its point or points of equilibrium, considering the cases $c < 0, c = 0, c > 0$, separately.

3. The dynamics of a nonlinear system are expressed by

$$\left.\begin{aligned} \dot{x}_1 &= ax_1 - x_2 - bx_1^3 \\ \dot{x}_2 &= ax_2 + x_1 - bx_2^3 \end{aligned}\right\} \qquad (a, b > 0)$$

Show (i) analytically or graphically that there are in general nine equilibrium points, such that the origin is always one such point, and such that the other eight are either all present or all absent and may, when present, be divided into two groups of four, each group having co-ordinates (X_1, X_2), $(-X_2, X_1)$, $(-X_1, -X_2)$ and $(X_2, -X_1)$

(ii) that the origin is always unstable

(iii) that if $a = 17/6$ and $b = 5/6$, there is an equilibrium point at $(1, 2)$; deduce the location of the other equilibrium point in the first quadrant and investigate the stability of each

(iv) by using the Liapunov function $V = x_1^2 + x_2^2$, that if a trajectory starts at a sufficiently large distance from the origin, there is always a smaller radius outside which the trajectory approaches the origin monotonically.

4. A continuous-flow, stirred-tank chemical reaction obeys the equations:

$$\dot{S} = a(S_0 - S) - kS\exp(-c/T),\ \dot{T} = a(T_0 - T) + bkS\exp(-c/T)$$

where

S = concentration of reactant in tank and effluent
S_0 = concentration of reactant in influent
T = temperature of tank contents and effluent
T_0 = temperature of influent
a, b, c, k = positive constants

Write down the equilibrium conditions for the process and deduce that for any equilibrium state, T satisfies the equation

$$1 + \frac{a}{k}\exp(c/T) = \frac{bS_0}{T - T_0}$$

Show that if $T_0 = 300°\text{C}$, $S_0 = 10$, $a = 2^{-9}$, $b = 30$, $k = 1/2$ and $c = 3600 \log_e 2$, this equation is satisfied by $T = 360°\text{C}$, $400°\text{C}$ and $450°\text{C}$ and deduce the associated values of S. Linearise the state equations in the neighbourhood of each equilibrium point and show that only two of these points are stable in the small.

Part 2

Some techniques of feedback controller design

Chapter 5

Concepts of feedback control

5.1 Introduction

Part I of this book dealt with some essential properties of systems in general. In Part II we shall be dealing with feedback control systems and considering some of the methods which have been suggested for their design. All the methods outlined are restricted to systems which have a linear, autonomous representation. The majority of design methods in this field tend to use complex-frequency system analysis rather than time domain methods, though, as seen in earlier Chapters, a state-vector representation in the time domain is easily converted to a transfer matrix representation and, though perhaps not so easily, *vice versa*.

5.1.1 Basic control system

The general lay-out of a feedback control system is shown in Fig. 5.1 in Laplace notation.

To avoid over-using the term 'system', the original system to which it is required to apply control will be called the *plant*, and its transfer matrix is denoted by $\boldsymbol{P}(s)$. It will be noted that the input to and the output from the plant are, as in the general analysis of Part I, denoted by the vectors $\boldsymbol{U}(s)$, of order m, and $\boldsymbol{V}(s)$, of order p, respectively.

The elements of $\boldsymbol{V}(s)$, namely the quantities to be controlled, are fed back through the *feedback matrix* $\boldsymbol{H}(s)$ and the output from $\boldsymbol{H}(s)$ is compared with the external *requirement vector* $\boldsymbol{R}(s)$, the difference between these vectors being the *error vector* $\boldsymbol{E}(s)$. [Strictly speaking $\boldsymbol{E}(s)$ is only a true error vector if $\boldsymbol{H} = \boldsymbol{I}_p$, which is not necessarily the case.] To complete the loop, $\boldsymbol{E}(s)$ is fed into the *controller matrix* $\boldsymbol{K}(s)$,

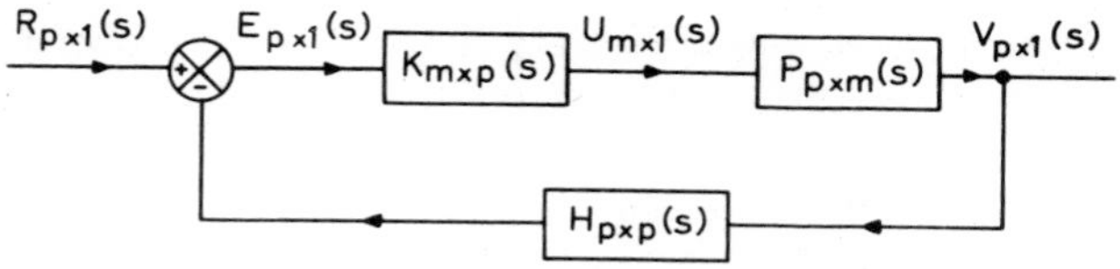

Fig. 5.1 Typical feedback control system

the output from the controller being the input $U(s)$ to the plant, which, it is hoped, will modify the output $V(s)$ in such a way that the error will be reduced. It has been assumed in the diagram that the order of the requirement vector, $\boldsymbol{R}(s)$, is the same as that of the output vector, $V(s)$: this is in general, though not always, the case since the purpose of the control operation is usually to place a requirement on each element of the output.

Essentially, the design of the control system implies finding values of $\boldsymbol{K}(s)$ and $\boldsymbol{H}(s)$ such that the behaviour of the control system complies with certain requirements, the more important of which we now outline.

5.1.2 Desirable properties of control systems

(i) In the first place it is *essential* that the control system should be stable, (even in those rare cases in which the plant itself is unstable).
(ii) The system usually has to meet accuracy requirements, in the sense that, ultimately at any rate, the elements of the output vector $\boldsymbol{v}(t)$ should approximate to their requirement values within a prescribed tolerance.
(iii) The approach to the ultimate value should be reasonably rapid and free from excessive overshoot: these requirements are clearly related to the *degree* of stability of the system.
(iv) It is normally desirable to limit the degree of *interaction* or *crosstalk* in the system. In general, depending on the structure of the system transfer matrix $\boldsymbol{F}(s)$ relating $V(s)$ to $\boldsymbol{R}(s)$, any one element of $\boldsymbol{R}(s)$ will affect *every* element of $V(s)$ instead of only one: this is what is meant by interaction. It is desirable to minimise interaction, which implies making $\boldsymbol{F}(s)$ approximate to a diagonal form, for a variety of reasons: for instance because undue interaction may reduce the stability margins of system operation (i.e. the permissible gains in the various loops present in the system); from a design point of view, moreover, several of the design techniques to be presented seek to reduce the design problem to that of p uncoupled (noninteracting) loops, in order that the well-tried design methods used for one-input/one-output systems may be applied to each such loop.

(v) It is often required that the system should have the property of *integrity*: by this is meant that the system should remain stable even if certain failures occur in the system, notably if a break should occur in one or more of the feedback paths.
(vi) Preferably, the structure of the feedback and controller units should be as simple as possible, both in mathematical and in physical terms.
(vii) Finally, the system should be reasonably immune from the effects of disturbances or noise, whether internally developed or externally injected: this topic will only be touched upon in the present text which is limited to deterministic systems.

It will be clear that, as usual in design problems, a satisfactory design will be a compromise solution between possibly conflicting requirements: for instance high accuracy demands high gains which tend to produce instability, the complete elimination of interaction often leads to unnecessarily complex controller structures etc.

5.2 Basic relations and some useful concepts

We deduce immediately from Fig. 5.1:

$$V(s) = P(s)U(s); \quad U(s) = K(s)E(s);$$

Hence

$$V(s) = P(s)K(s)E(s) \equiv G(s)E(s)$$

where

$$G(s)_{p\times p} = P(s)_{p\times m}K(s)_{m\times p} = \text{forward transmittance matrix} \tag{5.1}$$

(Note that $G = PK$ and not KP: the matrix product order runs counter to the causatory chain order.)

Also

$$R(s) = E(s) + H(s)V(s)$$

so that

$$\begin{aligned} G(s)R(s) &= G(s)E(s) + G(s)H(s)V(s) \\ &= [I_p + G(s)H(s)]\,V(s) \end{aligned}$$

and hence

$$\begin{aligned} V(s) &= [I_p + G(s)H(s)]^{-1}G(s)R(s) \\ &\equiv F(s)R(s) \end{aligned}$$

where

$$\begin{aligned} F(s) &= [I_p + G(s)H(s)]^{-1}G(s) \\ &= \text{closed-loop transfer matrix} \end{aligned} \tag{5.2a}$$

Note that, provided only $G(s)$ is nonsingular, we may write

$$F(s) = [I_p + G(s)H(s)]^{-1}[G^{-1}(s)]^{-1}$$
$$= [G^{-1}(s) + H(s)]^{-1}$$

so that

$$F^{-1}(s) = G^{-1}(s) + H(s) \tag{5.2b}$$

5.2.1 The loop-gain matrix

Departing temporarily from Fig. 5.1, consider a more general loop consisting, in causatory order, of any number of transfer matrices forming a re-entrant, conformable (under multiplication) sequence, say $\boldsymbol{A}_{b\times a}$, $\boldsymbol{B}_{c\times b}, \ldots, \boldsymbol{G}_{h\times g}, \boldsymbol{H}_{a\times h}$. Imagine this loop broken at any point, say between $\boldsymbol{C}$ and $\boldsymbol{D}$. Feed an input vector $\boldsymbol{\theta}(s)$ of order d into $\boldsymbol{D}$. The output from $\boldsymbol{C}$ will be $\boldsymbol{C}_{d\times c}(s).\boldsymbol{B}_{c\times b}(s) \ldots \boldsymbol{D}_{e\times d}(s).\boldsymbol{\theta}(s)$ or, more briefly, $(\boldsymbol{CBAHGFED})_{d\times d}\boldsymbol{\theta}$. We define the loop-gain matrix for this particular break-point by

$$\boldsymbol{L}_{CD}(s) = (\boldsymbol{CBAHGFED})_{d\times d}$$

Note that the loop-gain matrix is necessarily square but that its order may depend upon the break-point. Even if all the separate matrices are square and of the same order, $\boldsymbol{L}$ will still depend upon the break-point for its value unless $\boldsymbol{A}, \boldsymbol{B}, \ldots, \boldsymbol{H}$ commute under multiplication.

Moreover, if the lowest number in the set of positive integers $a, b, \ldots, h$ is denoted by z, then if $d > z$, $\boldsymbol{L}_{CD}$ is singular. For, by supposition, in the product $\boldsymbol{CB} \ldots \boldsymbol{D}$ there is at least one factor with z rows and another with z columns so that the rank of this product cannot exceed z. If therefore $d > z$, it follows that the rank of $\boldsymbol{L}_{CD}$ is smaller than its order, so that $\boldsymbol{L}_{CD}$ is singular.

5.2.2 The return-difference matrix

With the same loop as above, the same break-point between $\boldsymbol{C}$ and $\boldsymbol{D}$ and the same input, $\boldsymbol{\theta}$, to $\boldsymbol{D}$, the *difference* between this input and the output from $\boldsymbol{C}$ which it creates is clearly $(\boldsymbol{I}_d - \boldsymbol{CB} \ldots \boldsymbol{ED})\boldsymbol{\theta}$. We define the return-difference matrix for this break-point by

$$\boldsymbol{\Delta}_{CD} = \boldsymbol{I}_d - \boldsymbol{CB} \ldots \boldsymbol{ED} = \boldsymbol{I}_d - \boldsymbol{L}_{CD}$$

Clearly $\boldsymbol{\Delta}_{CD}$ suffers the same variations with break-point as $\boldsymbol{L}_{CD}$, *but* it has the important property that det $\boldsymbol{\Delta}_{CD}$ is independent of the break-point. Indeed, consider the values of $\boldsymbol{\Delta}$ for any two break-points, say

$$\boldsymbol{\Delta}_{CD} = \boldsymbol{I}_d - \boldsymbol{CBAHGFED} \quad \text{and} \quad \boldsymbol{\Delta}_{FG} = \boldsymbol{I}_g - \boldsymbol{FEDCBAHG}$$

By established matrix theory, the value of the determinant of the partitioned matrix $\begin{bmatrix} I_d & (CBAHG)_{d\times g} \\ (FED)_{g\times d} & I_g \end{bmatrix}$ may be written as

either

$$\det I_d . \det (I_g - FED . I_d^{-1} . CBAHG) = \det (I_g - FEDCBAHG) = \det \mathbf{\Delta}_{FG}$$

or

$$\det I_g . \det (I_d - CBAHG . I_g^{-1} . FED) = \det (I_d - CBAHGFED) = \det \mathbf{\Delta}_{CD}$$

Thus det $\mathbf{\Delta}_{CD}$ = det $\mathbf{\Delta}_{FG}$ for an arbitrarily chosen pair of break-points. Hence det $\mathbf{\Delta}$ is independent of the break-point even though the *order* of $\mathbf{\Delta}$ in general varies with the break-point.

5.2.3 Application to Fig. 5.1

The loop here consists effectively of K, P and $-H$ in causatory order. We deduce

$$L_{KP} = K(-H)P = -KHP; L_{P(-H)} = PK(-H) = -PKH;$$
$$L_{(-H)K} = (-H)PK = -HPK \tag{5.3}$$

these matrices being square and of order m, p and p respectively. (Some writers, we feel rather illogically, define the loop-gain matrices as the negatives of the above quantities.) Moreover

$$\mathbf{\Delta}_{KP} = I_m + KHP; \mathbf{\Delta}_{P(-H)} = I_p + PKH; \mathbf{\Delta}_{(-H)K} = I_p + HPK \tag{5.4}$$

these return-difference matrices being also square, of order m, p and p respectively, but such that their determinants are identical in value.

5.3 State-vector analysis

In translating the system representation from state vector form to transfer matrix form, or *vice versa*, we stress the following points already made in Part I. The transfer matrix representation excludes the possible presence of uncontrollable or unobservable modes: it may therefore be a fruitless exercise to carry out a stability study of a system based upon its transfer matrix if any such modes exist and are either unstable or even lightly damped. However, if the *only* information we have about a system is its transfer matrix, then we have no

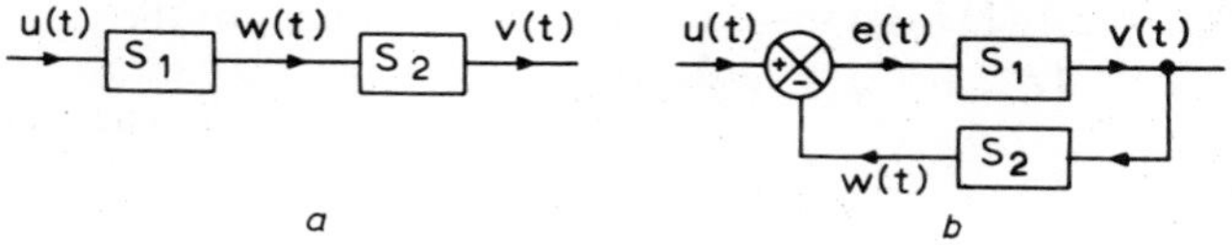

Fig. 5.2 Interconnection of two subsystems
(a) series connection
(b) feedback connection

information about uncontrollable and unobservable modes and in translating the transfer matrix into a state-vector representation, the latter should be of minimal order. With this preamble, and assuming that each element in the loop (K, P and H) is capable of state-vector representation, we propose to analyse the state-vector representation of such elements both in series (K and P) and in feedback formation (G and $-H$). (Note that if a transfer matrix $T(s)$ is to be capable of state-vector representation then $\operatorname*{Lt}_{s\to\infty} T(s)$ must be a finite matrix, for this limit, as pointed out in Section 3.4.1, is the value of D. This condition, although usually met, is *not* met by the feedback matrix $H(s)$ in, for instance, a position-control system in which velocity feedback is used, for $H(s)$ will then contain a term linear in s.) For simplicity of notation we consider two subsystems, S_1 and S_2, representable in state-vector form by (A_1, B_1, C_1, D_1) and (A_2, B_2, C_2, D_2) respectively. Let the state-vectors be $x_1(t)$, of order n_1, and $x_2(t)$, of order n_2, respectively, and let the transfer matrices be $T_1(s)$, $T_2(s)$.

5.3.1 Series connection (Fig. 5.2a)

By supposition the following equations hold:

$$\dot{x}_1 = A_1x_1 + B_1u \quad \dot{x}_2 = A_2x_2 + B_2w = A_2x_2 + B_2C_1x_1 + B_2D_1u$$
$$w = C_1x_1 + D_1u \quad v = C_2x_2 + D_2w = C_2x_2 + D_2C_1x_1 + D_2D_1u$$

$$\text{Hence, writing } x = \begin{bmatrix} x_1 \\ x_2 \end{bmatrix}, \dot{x} = \begin{bmatrix} A_1 & O \\ B_2C_1 & A_2 \end{bmatrix} x + \begin{bmatrix} B_1 \\ B_2D_1 \end{bmatrix} u \equiv Ax + Bu, \text{ say,}$$

$$\text{and } v = [D_2C_1, C_2]\, x + D_2D_1u \equiv Cx + Du, \text{ say,} \tag{5.5}$$

which are the state vector equations of the series combination in terms of a state vector x of order $(n_1 + n_2)$. The transfer matrix of the combination is of course

$$T(s) = T_2(s)T_1(s)$$
$$= \{C_2(sI-A_2)^{-1}B_2+D_2\}\{C_1(sI-A_1)^{-1}B_1+D_1\} \tag{5.6}$$

5.3.2 Feedback connection (Fig. 5.2*b*)

In this case we have the five equations:

$$\dot{x}_1 = A_1x_1+B_1e; \quad \dot{x}_2 = A_2x_2+B_2v; \quad e = u-w$$
$$v = C_1x_1+D_1e; \quad w = C_2x_2+D_2v;$$

Substituting for w from the fourth equation and then for e from the fifth leads to:

$$\dot{x}_1 = A_1x_1-B_1C_2x_2+B_1u-B_1D_2v;$$
$$v = C_1x_1-D_1C_2x_2+D_1u-D_1D_2v; \quad \dot{x}_2 = A_2x_2+B_2v$$

Assuming that $(I+D_1D_2)$ is nonsingular and writing for brevity $(I+D_1D_2)^{-1} \equiv K$, the second equation yields

$$v = K(C_1x_1-D_1C_2x_2+D_1u)$$

which, substituted in the first and third, leads to:

$$\dot{x}_1 = (A_1-B_1D_2KC_1)x_1-(B_1C_2-B_1D_2KD_1C_2)x_2$$
$$+(B_1-B_1D_2KD_1)u$$
$$\dot{x}_2 = B_2KC_1x_1+(A_2-B_2KD_1C_2)x_2+B_2KD_1u$$

Hence, using again the compound state vector $x = \begin{bmatrix} x_1 \\ x_2 \end{bmatrix}$ the effective matrix parameters of the feedback combination are:

$$A = \begin{bmatrix} A_1-B_1D_2KC_1 & B_1(D_2KD_1-I)C_2 \\ B_2KC_1 & A_2-B_2KD_1C_2 \end{bmatrix}; \quad B = \begin{bmatrix} B_1(I-D_2KD_1) \\ B_2KD_1 \end{bmatrix};$$
$$C = [KC_1, -KD_1C_2]; \quad D = KD_1 \tag{5.7}$$

The effective transfer matrix is in this case $T(s) = \{I+T_1(s)T_2(s)\}^{-1}T_1(s)$ and since it has been assumed that $T_1(s)$ and $T_2(s)$ have finite limits D_1 and D_2 as $s \to \infty$, the corresponding limit of $T(s)$ is $(I+D_1D_2)^{-1}D_1 = KD_1$. Thus the feedback combination is only state-representable if K is a finite matrix, i.e. if $(I+D_1D_2)$ is nonsingular, which is the assumption made above.

5.3.3 Some comments

If $T_1(s)$ is numeric (i.e. not a function of s) then its state representation is contained entirely in its output equation, of the form $v = T_1 u$: there is *no* dynamic state-equation, or, in other words, the order of the state-vector is zero. The associated A_1 matrix has therefore no rows and no columns, B_1 has no rows, C_1 has no columns, but $D_1 = T_1$. The same argument holds for $T_2(s)$. Such a situation of course considerably simplifies the more general results given in eqns. 5.5 and 5.7.

It is to be noted that even if the state-vector forms of $T_1(s)$ and $T_2(s)$ are minimal, this is only a necessary and not a sufficient condition that the overall state-vector representation, whether in the series case or in the feedback case, should be minimal. However, if the overall representation is *not* minimal, any noncontrollable or nonobservable modes which it contains are not fictitious but are physically present in the series or feedback combination considered, and should therefore *not* be eliminated by seeking a minimal representation.

From a design point of view, however, the hard fact must be faced that the state-vector representation of the three elements K, P and H in Fig. 5.1 is *not* a fruitful approach. In the theory given above we have not gone so far as to consider the matrix parameters resulting from two subsystems (K and P) in series, bridged by a further subsystem, $(-H)$, as feedback. The complexity of such results is obvious without obtaining them. Moreover it is very difficult to relate the overall parameters of a feedback system to the various criteria of performance mentioned in Section 5.1.2, with the possible exceptions of stability and accuracy, quite apart from the problem of disentangling the separate effects of the matrix parameters of $K(s)$ and $H(s)$ upon the overall parameters. Whatever uses the above analysis has it is *not* in the field of design.

5.4 The modal polynomial

The modal polynomial of a system may be defined as a monic polynomial in s, the zeros of which determine the natural modes present in the state elements, in the sense that a zero $s = a$ determines a mode $\exp(at)$, a double zero $s = a$ determines modes $\exp(at)$ and $t.\exp(at)$ etc., every such mode being present in at least one of the state elements when the system is unexcited but subject to *arbitrary* initial conditions $x(0) = x_0$. Using the results of Section 2.5, under these conditions

$$X(s) = (sI - A)^{-1} x_0 = \frac{\operatorname{adj}(sI - A)}{\det(sI - A)} x_0 = \operatorname{adj}(sI - A).x_0 \Big/ \prod_1^n (s - \lambda_i)$$

When every element of $X(s)$ is broken up into partial fractions and the inverse Laplace transform is taken, it follows that the modes present in the various elements of $\boldsymbol{x}(t)$ must all be of the forms: $t^j \exp(\lambda_i t)$, $j = 0, 1, 2, \ldots, m_i - 1$; $i = 1, 2, \ldots, d$; where λ_i is an eigenvalue of A of multiplicity m_i and d is the number of *distinct* eigenvalues. Some of these modes may be absent because a particular value of $\boldsymbol{x}_0$ is chosen which makes the coefficients of a particular mode zero. Some modes may also be absent (whatever the value of $\boldsymbol{x}_0$) if some factor $(s - \lambda_i)$ of $\det(s\boldsymbol{I} - A)$ is also a factor of $\operatorname{adj}(s\boldsymbol{I} - A)$; it may be shown that this situation can only occur if the relevant λ_i is a *multiple* eigenvalue (and may not occur even then) and in any case the factor $(s - \lambda_i)$ will always occur to a higher power in $\det(sI - A)$ than in $\operatorname{adj}(sI - A)$, so that, after cancellation, the factor is still present in the denominator; since from a stability point of view we are mainly interested in the *location* of the eigenvalue rather than its multiplicity, this partial cancellation is therefore trivial. We are therefore justified in saying that, *for a linear, autonomous system expressed in state vector form the modal polynomial is* $\det(s\boldsymbol{I} - A)$.

5.4.1 Broken-loop and closed-loop modal polynomials

Consider a broken-loop sequence for which the overall transfer matrix is $\boldsymbol{L}(s)$, so that if the input to the sequence is $\boldsymbol{\theta}(s)$, the output will be $\boldsymbol{L}(s).\boldsymbol{\theta}(s) \equiv \boldsymbol{\Phi}(s)$, say. Let $\boldsymbol{L}(\infty)$ be finite (zero or not) so that the sequence has a state-vector representation, say in the Laplace domain:

$$sX(s) = AX(s) + \boldsymbol{B}\boldsymbol{\theta}(s); \quad \boldsymbol{\Phi}(s) = CX(s) + \boldsymbol{D}\boldsymbol{\theta}(s)$$

Then, by the previous section, the broken-loop modal polynomial (b.l.m.p.) is given by

$$\text{b.l.m.p.} = \det(s\boldsymbol{I} - A) \tag{5.8}$$

Now let us close the loop by bonding the output to the input, i.e. by making $\boldsymbol{\Phi}(s)$ equal to $\boldsymbol{\theta}(s)$. The previous output equation becomes

$$(\boldsymbol{I} - \boldsymbol{D})\boldsymbol{\theta}(s) = CX(s)$$

or, provided only that $(\boldsymbol{I} - \boldsymbol{D})$ is nonsingular, $\boldsymbol{\theta}(s) = (\boldsymbol{I} - \boldsymbol{D})^{-1}CX(s)$. Substituting this value in the state equation leads to

$$sX(s) = \{A + \boldsymbol{B}(\boldsymbol{I} - \boldsymbol{D})^{-1}C\}X(s)$$

a state equation in which the equivalent 'A' matrix is $\{A + \boldsymbol{B}(\boldsymbol{I} - \boldsymbol{D})^{-1}C\}$. Hence the closed-loop modal polynomial (c.l.m.p.) is given by

$$\text{c.l.m.p.} = \det\{sI - A - B(I-D)^{-1}C\}$$
$$= \det(sI - A).\det\{I - (sI-A)^{-1}B(I-D)^{-1}C\}$$

Hence $\dfrac{\text{c.l.m.p.}}{\text{b.l.m.p.}} = \det\{I - C(sI-A)^{-1}B(I-D)^{-1}\}$, (using Section 5.2.2)

$$= \det\{I - D - C(sI-A)^{-1}B\}/\det(I-D)$$
$$= \det\{I - L(s)\}/\det\{I - L(\infty)\}$$
$$= \det \mathbf{\Delta}(s)/\det \mathbf{\Delta}(\infty) \qquad (5.9)$$

since on the one hand $L(s) = C(sI-A)^{-1}B + D$ by eqn. 2.9, while on the other, the return difference matrix $\mathbf{\Delta}(s) = I -$ (loop-gain matrix) $= I - L(s)$ from Section 5.2.2. The simple result of eqn. 5.9 is widely used in control-system design.

5.4.1.1 Notes

(i) The denominator of eqn. 5.9, det $\mathbf{\Delta}(\infty)$, is a pure number

(ii) Since, from Section 5.2.2, det $\mathbf{\Delta}(s)$ is independent of the break-point, so is the result given by eqn. 5.9.

(iii) Det $\mathbf{\Delta}(s)$ is normally a rational function of s, namely one polynomial in s divided by another; it is therefore tempting to identify the numerator polynomial with the c.l.m.p. and the denominator with the b.l.m.p. (apart from a possible numerical factor due to det $\mathbf{\Delta}(\infty)$). It may be, however, that the b.l.m.p. and the c.l.m.p. have a common factor, in which case this identification is of course erroneous if det $\mathbf{\Delta}(s)$ is expressed in its lowest terms; if, however, det $\mathbf{\Delta}(s)$ is evaluated *without any cancellations taking place*, this identification is still permissible.

(iv) It follows from (iii) that the criterion of stability of the closed-loop system is that det $\mathbf{\Delta}(s)$, *evaluated without cancellations*, must have no zeros in the closed right-half plane of s.

5.5 Conclusion

We have not, in this Chapter, considered any methods of design: we have only introduced certain concepts relating to feedback systems which underlie some of the methods of design considered in the remaining chapters of Part II. The emphasis, not unnaturally, has been in relation to the problem of stability.

5.6 References

MACFARLANE, A.G.J.: 'Return-difference and return-ratio matrices and their use in analysis and design of multivariable feedback control systems', *Proc. IEE*, 1970, **117**, (10), pp. 2037–49

5.7 Examples 5

1.

$$\text{If } \boldsymbol{A} = \begin{bmatrix} 1 & 1 & 0 & 2 \\ 2 & 0 & 1 & 3 \\ 1 & 1 & 1 & 1 \end{bmatrix}, \boldsymbol{B} = \begin{bmatrix} 1 & -1 & 2 \\ 3 & 1 & 2 \\ 1 & 1 & 2 \end{bmatrix}, \boldsymbol{C} = \begin{bmatrix} 1 & 4 & 1 \\ 2 & 1 & 0 \end{bmatrix}, \boldsymbol{D} = \begin{bmatrix} 2 & 1 \\ 3 & 2 \\ 4 & 1 \\ 1 & 2 \end{bmatrix},$$

evaluate the products $\boldsymbol{DCBA}$, $\boldsymbol{CBAD}$, $\boldsymbol{BADC}$ and $\boldsymbol{ADCB}$. Show that all these matrices are of rank 2.

2. In Fig. 5.1, with $\boldsymbol{G} = \boldsymbol{PK}$, let

$$\boldsymbol{G} = \begin{bmatrix} k_a/(s+a) & k_b/(s+b) \\ k_c/(s+c) & k_d/(s+d) \end{bmatrix} \text{ and let } \boldsymbol{H} = \begin{bmatrix} k_A/(s+A) & k_B/(s+B) \\ k_C/(s+C) & k_D/(s+D) \end{bmatrix}$$

Evaluate in general the transfer matrices $\boldsymbol{GH}$, $\boldsymbol{HG}$, $(\boldsymbol{I} + \boldsymbol{GH})^{-1}\boldsymbol{G}$ and the scalar quantity $\det(\boldsymbol{I} + \boldsymbol{GH}) = \det(\boldsymbol{I} + \boldsymbol{HG})$. Show that *in general* the ratio of the closed-loop modal polynomial (c.l.m.p.) to the broken-loop modal polynomial (b.l.m.p.) is equal to $\det(\boldsymbol{I} + \boldsymbol{GH}) \equiv \det \boldsymbol{\Delta}$.

Is this statement still valid if (i) $b = a$; (ii) $d = a$; (iii) $d = a$ *and* $C = B$; (iv) $C = B = d = a$? If not, what is the modified relation? [Note: the tedious algebra is considerably curtailed by the use of an abbreviated notation, e.g.: $(s + a) = s_a$; $s_a s_b \ldots = s_{ab\ldots}$; $k_a/(s + a) = (a)$; $(a).(b) = (ab)$ etc.]

3. In Example 2, as an instance of case (iv), take $(a, b, c, d) = (1, 2, 5, 1)$ and $(A, B, C, D) = (3, 1, 1, 4)$, leaving the k-parameters arbitrary. Find the closed-loop modal polynomial by expressing the subsystems $\boldsymbol{G}(s)$ and $\boldsymbol{H}(s)$ in state-vector form and then using the results of Section 5.3.2.

4. In Fig. 5.1 take $\boldsymbol{P} = \dfrac{1}{(s+2)(s+3)} \begin{bmatrix} s-1 & s+2 \\ s-3 & s+3 \end{bmatrix}$, $\boldsymbol{K} = \dfrac{1}{s+4} \begin{bmatrix} s+2 & s+3 \\ s+3 & s+5 \end{bmatrix}$ and $\boldsymbol{H} = \dfrac{1}{s+1} \begin{bmatrix} 1 & 2 \\ 2 & 3 \end{bmatrix}$. Evaluate the products $\boldsymbol{PK}$, $\boldsymbol{KH}$, $\boldsymbol{HP}$, $\boldsymbol{PKH}$, $\boldsymbol{KHP}$, $\boldsymbol{HPK}$ in the form of a polynomial matrix divided by a

scalar polynomial, cancelling any common factors. Verify that det (return difference matrix) is independent of the break-point.

Taking the break-point between $\boldsymbol{P}$ and $\boldsymbol{H}$, obtain the ratio of the closed-loop modal polynomial (for the transfer matrix relating $\boldsymbol{V}$ to $\boldsymbol{R}$) to the broken-loop modal polynomials for (i) the overall transfer matrix $\boldsymbol{PKH}$ only, (ii) $\boldsymbol{PKH}$ as well as the interim transfer matrices $\boldsymbol{KH}$ and $\boldsymbol{H}$. Is the relation of eqn. 5.9 satisfied in either case? Is the closed-loop system stable?

Chapter 6

On pole location

6.1 Poles and zeros of transfer matrices

The transfer matrix of a unit is the matrix relating the Laplace transform of the output vector of the unit to that of the input vector to the unit. Its elements are normally rational functions of s (i.e. a polynomial in s divided by another) and the degree of the numerator polynomial is normally smaller than, or at most equal to, that of the denominator polynomial. (We here exclude from consideration units containing time delays or transportation lags, for which the elements contain exponential functions of s of the form $\exp(-sT)$; note in passing that, strictly speaking, every system contains transportation lags since it always takes a finite time for a signal to travel from one point to another: this is another example of approximation by mathematical modelling!)

Denoting such a matrix by $\boldsymbol{T}(s)$, any element may be written as $t_{ij}(s) = n_{ij}(s)/d_{ij}(s)$ where n and d are polynomials in s having no common factor. The zeros of $n_{ij}(s)$ are the zeros of the element $t_{ij}(s)$; the zeros of $d_{ij}(s)$ are its poles. If we denote the lowest common multiple of the several $d_{ij}(s)$ by $p_T(s)$, then $p_T(s)\,.\,\boldsymbol{T}(s)$ will be a polynomial matrix, denoted here by $\boldsymbol{P}_T(s)$ i.e. a matrix of which every element is a polynomial in s; moreover no polynomial of lower degree than $p_T(s)$ can achieve this result. The zeros of $p_T(s)$ i.e. values of s satisfying $p_T(s) = 0$, are defined as the *poles of the matrix* $\boldsymbol{T}(s)$.

Let the rank of $\boldsymbol{T}(s)$, considered as a function of an unspecified s, be R_T. Then any value of s which, when substituted in $\boldsymbol{T}(s)$, gives a numeric matrix of rank smaller than R_T is defined as a *zero* of $\boldsymbol{T}(s)$. (Note that whereas the poles of $\boldsymbol{T}(s)$ include the poles of every element, the zeros of $\boldsymbol{T}(s)$ are not, in general, related to the zeros of the elements.)

If $\boldsymbol{T}(s)$ is square and nonsingular (i.e. if $\det \boldsymbol{T}(s) \not\equiv 0$) then any zero of $T(s)$ reduces its rank and is therefore a zero of det $T(s)$ and therefore also a zero of det $P_T(s)$, though det $P_T(s)$ will have other zeros.

The emphasis in this Chapter will be on the poles of the closed-loop transfer matrix $F(s)$, since these control the stability of the closed-loop system; in fact $p_F(s)$ is the closed-loop modal polynomial of Chapter 5, but necessarily excluding any noncontrollable or nonobservable modes which may be present, since it is derived from a transfer matrix relationship.

6.2 The problem of pole location

Since 1967 a large number of papers have been published on this topic. The basic problem may be stated as follows: given a plant expressible in state vector form by the standard equations

$$\dot{x} = Ax + Bu, \quad v = Cx$$

design, if possible, a *numeric* feedback matrix $(-H)_{m\times n}$ from x to u so that the poles of the resulting closed-loop transfer matrix lie in pre-assigned positions in the s-plane. Alternatively, if all the elements of x are not accessible for feedback purposes, design, if possible, a numeric feedback matrix $(-H)_{m\times p}$ from v to u so that the poles of the resulting closed-loop transfer matrix lie in pre-assigned positions in the s-plane. (See Fig. 6.1)

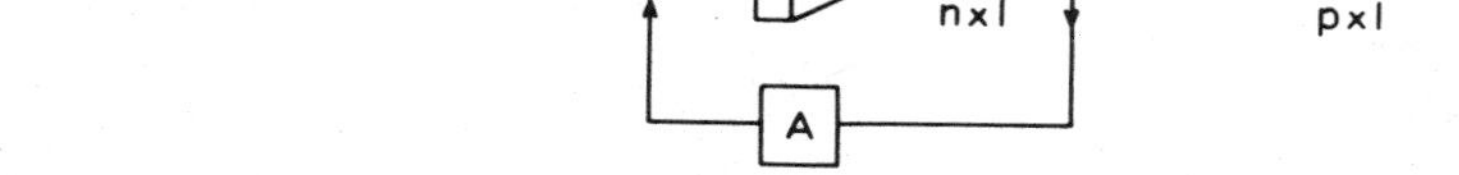

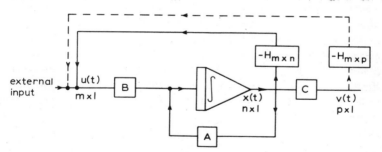

Fig. 6.1 Schematic of plant with state (full line) feedback, or output (broken line) feedback, to input

6.2.1 Comments on the problem

Before going further it is essential to appreciate that locating the poles of the closed-loop transfer matrix $F(s)$ achieves nothing, from a system performance point of view, other than determining the nature of the natural modes. The amplitudes of these natural modes in the various elements of the output, the accuracy of the system, the problem of interaction etc. are all affected by the nature of the matrix $P_F(s)$ whereas pole location only serves to fix the zeros of the polynomial $p_F(s)$. Although, where possible, we shall therefore try to go further than mere pole location, we begin by an exposition of its theory as presented in some of the leading papers on the subject.

Note that the problem, as stated, envisages the possibility of feedback from the state vector $\boldsymbol{x}$ rather than from the output vector $\boldsymbol{v}$ as in Fig. 5.1. This greater degree of freedom is not usually possible directly but may, under certain conditions, be achieved indirectly through so-called Luenberger observers, the function of which is to simulate the elements of $\boldsymbol{x}$ for feedback purposes from a knowledge of the accessible output $\boldsymbol{v}$ and the known input $\boldsymbol{u}$. The theory of Luenberger observers is outlined in Section 6.5.

Note also that the output equation normally associated with this problem assumes $\boldsymbol{D} = 0$. Whereas this is often the case for a plant, it is a minor limitation on the theory when output feedback as opposed to state feedback is used.

6.3 State feedback

The first major paper on the subject appears to be by Wonham;[1] an earlier paper by Luenberger[2] on canonical forms states, without proof, '. . . it can be shown that by linear state-vector feedback the poles of a multivariable system can be placed arbitrarily'. Wonham's proof is somewhat sophisticated mathematically and we prefer to follow Chen,[3] with the simplifying assumption that the eigenvalues of A are distinct.

6.3.1

We require as a preliminary the *phase-variable canonical form* of the state equation. This form of the $\boldsymbol{A}$-matrix has already been developed in Section 1.6. The corresponding form of the $\boldsymbol{B}$-matrix is a matrix of order $n \times m$ of which the first column is null except for its last element which is unity.

It was shown in Section 1.6 that if we start from the diagonal canonical form

$$\dot{y} = \Lambda y + \boldsymbol{\beta} u \quad \text{where} \quad y = E^{-1}x, \Lambda = E^{-1}AE \quad \text{and} \quad \boldsymbol{\beta} = E^{-1}B$$

then a further nonsingular transformation $z = Py$ leads to

$$\dot{z} = P\Lambda P^{-1}z + P\boldsymbol{\beta}u \equiv Sz + P\boldsymbol{\beta}u$$

where

$$S = P\Lambda P^{-1} = PE^{-1}.A.EP^{-1}$$

and is of the particular form given in Section 1.6, namely

$$S = \begin{bmatrix} 0 & 1 & 0\ldots & 0 & 0 \\ 0 & 0 & 1\ldots & 0 & 0 \\ \cdot & \cdot & \cdots\cdot & \cdot & \cdot \\ 0 & 0 & 0\ldots & 0 & 1 \\ -a_n & -a_{n-1} & \ldots\ldots & -a_2 & -a_1 \end{bmatrix}$$

in which the a_r are the coefficients in the characteristic function of A, $\det(\lambda I - A) = \lambda^n + a_1\lambda^{n-1} + \ldots + a_{n-1}\lambda + a_n$, which, with s replacing λ, is the modal polynomial of the plant, $\det(sI - A)$. It was shown that, assuming A has distinct eigenvalues, P can always be found and is given by

$$P = V.\,\text{diag}\,p_{1r}$$

where V is a Vandermonde matrix of the eigenvalues and is nonsingular and where the p_{1r} are arbitrary but nonzero quantities.

We now require the further condition that the first column of $P\boldsymbol{\beta}$ should be $[0, 0, \ldots, 0, 1]'$. Now

$$\begin{aligned} \text{first column of } P\boldsymbol{\beta} &= P.(\text{first column of } \boldsymbol{\beta}) \\ &= V.\text{diag}\,p_{1r}.[\beta_{11}, \beta_{21}, \ldots, \beta_{n1}]' \\ &= V.[p_{11}\beta_{11}, p_{12}\beta_{21}, \ldots, p_{1n}\beta_{n1}]' \\ &= [0, 0, \ldots, 0, 1]' \text{ by supposition.} \end{aligned}$$

Hence

$$\begin{aligned} [p_{11}\beta_{11}, p_{12}\beta_{21}, \ldots, p_{1n}\beta_{n1}]' &= V^{-1}[0, 0, \ldots, 0, 1]' \\ &= \text{last column of } V^{-1} \end{aligned}$$

But the last column of V^{-1} consists of the cofactors of the last row of V, each divided by det V, and these cofactors are clearly, by inspection of V, the determinants of Vandermonde matrices of order $(n - 1)$ and therefore nonzero. Hence every product $p_{1r}\beta_{r1}$ $(r = 1, 2, \ldots n)$ must be equated to some known nonzero quantity. Hence, since the β_{r1} are known, the p_{1r} may be found uniquely, provided that none of the β_{r1} are zero, i.e. *provided* $\boldsymbol{\beta}$ *has no zero elements in its first column*.

Chen[3] overcomes this drastic constraint on $\boldsymbol{\beta}$ by applying a *nonsingular* transformation to the *input*, say $u = Mw$, so that

$P\boldsymbol{\beta}u$ = [illegible]. $\boldsymbol{\beta}M.w$ and $\boldsymbol{\beta}M$ therefore effectively replaces $\boldsymbol{\beta}$. Can we find M so t[illegible] M has no zero elements in its first column? Since the first column of $\boldsymbol{\beta}M$ equals $\boldsymbol{\beta}$. (first column of M), all that is required is that the scalar products (rth row of $\boldsymbol{\beta}$). (first column of M) shall be nonzero for $r = 1, 2, \ldots n$; in other words, the first column of M must *not* be orthogonal to any row of $\boldsymbol{\beta}$. This trivial constraint on M (its other columns are completely arbitrary,) can always be satisfied provided no row of $\boldsymbol{\beta}$ is null, i.e. *provided the plant is completely controllable*, but can clearly not be satisfied otherwise.

In summary then, if and only if (A, B) is a completely controllable pair, the state equation may be transformed to the phase-variable canonical form

$$\dot{z} = Sz + Tw \tag{6.1}$$

where $z = Py = PE^{-1}x$, $S = PE^{-1}AEP^{-1}$, $T = P\boldsymbol{\beta}M = PE^{-1}BM$, $w = M^{-1}u$, and where P, of the form $V.\operatorname{diag} p_{1r}$, can always be found, as well as M.

6.3.2 Introduction of unity-rank feedback

Suppose now that the input w is composed of an external input r together with a negative-feedback contribution from z, say $(-Jz)$, so that

$$w = r - Jz$$

Substitution in eqn. 6.1 gives

$$\dot{z} = (S - TJ)z + Tr$$

Suppose, moreover, that J is null except for its first row: then, bearing in mind the property of the first column of T, it follows that the product TJ is null except for its last row, which is precisely the first row of J. It then follows that $(S - TJ)$ has the same general form as S (given above) but that any last row element $(-a_r)$ in S becomes $-(a_r + j_{1,n+1-r})$ in $(S - TJ)$. In other words the plant modal polynomial

$$s^n + a_1 s^{n-1} + a_2 s^{n-2} + \ldots + a_{n-1}s + a_n$$

is changed by the feedback $(-J)$ to the closed-loop modal polynomial

$$\begin{aligned} s^n + (a_1 + j_{1n})s^{n-1} + (a_2 + j_{1,n-1})s^{n-2} \\ + \ldots + (a_{n-1} + j_{12})s + (a_n + j_{11}) \end{aligned} \tag{6.2}$$

Since, moreover, the j_{1r} may be chosen completely arbitrarily, the coefficients of this c.l.m.p. may be adjusted to any prescribed values, i.e., the poles of the closed-loop transfer matrix relating $\boldsymbol{z}$ to $\boldsymbol{r}$ may be assigned to arbitrary locations in the complex plane by a suitable choice of $\boldsymbol{J}$ (provided, of course, that any complex poles occur as conjugate pairs, since $\boldsymbol{J}$ is necessarily real).

Finally, the feedback equation $\boldsymbol{w} = \boldsymbol{r} - \boldsymbol{Jz}$ may be written

$$\boldsymbol{M}^{-1}\boldsymbol{u} = \boldsymbol{r} - \boldsymbol{JPE}^{-1}\boldsymbol{x} \quad \text{or} \quad \boldsymbol{u} = \boldsymbol{Mr} - \boldsymbol{MJPE}^{-1}.\boldsymbol{x}$$

showing that the feedback $(-\boldsymbol{J})$ from $\boldsymbol{z}$ to $\boldsymbol{w}$ is equivalent to a feedback $(-\boldsymbol{H}) = (-\boldsymbol{MJPE}^{-1})$ from $\boldsymbol{x}$ to $\boldsymbol{u}$ in the original representation of the plant, the external input being simultaneously changed from $\boldsymbol{r}$ to $\boldsymbol{Mr}$. Since $\boldsymbol{J}$, having only one non-null row, is of rank unity, so is $\boldsymbol{H} = \boldsymbol{MJPE}^{-1}$. Hence the following theorem:

Theorem 6.1 If a plant is completely controllable, a feedback matrix from state to input, of rank unity, may always be found such that the poles of the closed-loop transfer matrix may be assigned arbitrarily.

6.3.3 Comments on the theorem

(i) For a given plant and a given set of pre-assigned poles of the closed-loop transfer matrix, the j_{1r} $(r = 1, 2, \ldots n)$ are clearly uniquely determined by eqn. 6.2; so therefore is $\boldsymbol{J}$, whose other rows are null. $\boldsymbol{H}$, however, is *not* uniquely determined, for $\boldsymbol{H} = \boldsymbol{MJPE}^{-1}$ and $\boldsymbol{M}$ is not unique, as noted in the last section, though it must be nonsingular.
(ii) It may be shown that even when some eigenvalues of $\boldsymbol{A}$ are complex (and therefore occur as conjugate complex pairs) so that $\boldsymbol{E}$ is also complex, it is still always possible to find *real* values of $\boldsymbol{H}$.
(iii) Although the theorem establishes the *feasibility* of finding a suitable unity-rank feedback matrix, the actual determination of this matrix may be very laborious: it involves finding the eigenvalues of $\boldsymbol{A}$, an associated eigenvector matrix $\boldsymbol{E}$ as well as its inverse, possibly a suitable matrix $\boldsymbol{M}$, calculating $\boldsymbol{P}$, (which involves $\boldsymbol{V}$ and the last column of its inverse,) etc. Clearly, if m and n are at all large, these calculations, even computerised, are very tedious and any technique which can determine a suitable value of $\boldsymbol{H}$ more directly would be a great step forward.

6.3.4 An alternative approach

Such a technique is presented by Fallside and Seraji.[4] The inadequacy of pole location *per se* as a method of design is stressed, which reinforces the remarks in Section 6.2.1. In designing the feedback matrix to locate the closed-loop poles, it is therefore desirable to retain as many degrees of freedom as possible, in order to be able to modify, additionally, various aspects of performance other than stability. The paper nevertheless confines itself to feedback matrices of rank unity, due to the resulting simplicity of the analysis: since $\boldsymbol{H}$ is of order $m \times n$, however, and therefore has, in general, mn elements, the stipulation of unity rank reduces this number to $(m + n - 1)$, (say n arbitrary elements in any one row and $(m - 1)$ arbitrary multipliers of this row to give the other rows,) which is a fairly severe handicap.

6.3.4.1 Analysis of method

We start from the original representation of the plant

$$\dot{\boldsymbol{x}} \equiv \boldsymbol{A}\boldsymbol{x} + \boldsymbol{B}\boldsymbol{u} \quad \text{and} \quad \boldsymbol{v} = \boldsymbol{C}\boldsymbol{x}$$

in which it may be supposed, if required, that $\boldsymbol{B}$ and $\boldsymbol{C}$ are of maximal rank m and p respectively (see Section 1.7). Introducing a feedback matrix $(-\boldsymbol{H})$ from $\boldsymbol{x}$ to $\boldsymbol{u}$ and some external input $\boldsymbol{r}$ to $\boldsymbol{u}$, we deduce

$$\boldsymbol{u} = \boldsymbol{r} - \boldsymbol{H}\boldsymbol{x} \quad \text{so that} \quad \dot{\boldsymbol{x}} = (\boldsymbol{A} - \boldsymbol{B}\boldsymbol{H})\boldsymbol{x} + \boldsymbol{B}\boldsymbol{r}$$

Taking Laplace transforms with $\boldsymbol{x}(0) = \boldsymbol{0}$ and solving for $\boldsymbol{X}(s)$ gives

$$\begin{aligned} \boldsymbol{X}(s) &= (s\boldsymbol{I} - \boldsymbol{A} + \boldsymbol{B}\boldsymbol{H})^{-1}\boldsymbol{B}\boldsymbol{R}(s) \\ &= \operatorname{adj}(s\boldsymbol{I} - \boldsymbol{A} + \boldsymbol{B}\boldsymbol{H}).\boldsymbol{B}\boldsymbol{R}(s)/\det(s\boldsymbol{I} - \boldsymbol{A} + \boldsymbol{B}\boldsymbol{H}) \end{aligned}$$

the closed-loop modal polynomial being therefore $\det(s\boldsymbol{I} - \boldsymbol{A} + \boldsymbol{B}\boldsymbol{H})$.

Since $\boldsymbol{H}$ is of rank unity, it may be expressed as a (column).(row) product:

$$\boldsymbol{H} = \boldsymbol{q}_{m \times 1}.\boldsymbol{k}'_{1 \times n}$$

The c.l.m.p. becomes

$$\det(s\boldsymbol{I} - \boldsymbol{A} + \boldsymbol{B}\boldsymbol{q}\boldsymbol{k}') = \det(s\boldsymbol{I} - \boldsymbol{A}) + \boldsymbol{k}'.\operatorname{adj}(s\boldsymbol{I} - \boldsymbol{A}).\boldsymbol{B}\boldsymbol{q} \quad \text{(see Appendix)}$$

so that

$$\det(s\boldsymbol{I} - \boldsymbol{A} + \boldsymbol{B}\boldsymbol{q}\boldsymbol{k}') - \det(s\boldsymbol{I} - \boldsymbol{A}) = \boldsymbol{k}'.\operatorname{adj}(s\boldsymbol{I} - \boldsymbol{A}).\boldsymbol{B}\boldsymbol{q} \qquad (6.3)$$

The left hand side of this equation is the difference between the c.l.m.p. (of which the zeros are the prescribed closed-loop poles) and the plant modal polynomial, and is therefore a polynomial of degree $(n - 1)$, in general. *If* we now suppose that the elements of $\boldsymbol{q}$ have assigned values, then $\boldsymbol{Bq}$ is a known matrix and then $\boldsymbol{k}'$ and hence $\boldsymbol{H}$ may be found as follows.

If

$$\det(s\boldsymbol{I}-\boldsymbol{A}) = s_n + a_1 s^{n-1} + \ldots + a_{n-1}s + a_n$$

and

$$\det(s\boldsymbol{I}-\boldsymbol{A}+\boldsymbol{Bqk}') = s^n + c_1 s^{n-1} + \ldots + c_{n-1}s + c_n,$$

then eqn. 6.3 may be written

$$\begin{aligned}\boldsymbol{k}'.\mathrm{adj}(s\boldsymbol{I}-\boldsymbol{A}).\boldsymbol{Bq} &= (c_1-a_1)s^{n-1}+(c_2-a_2)s^{n-2}+\ldots+(c_n-a_n)\\ &= (\boldsymbol{c}'-\boldsymbol{a}')[s^{n-1}, s^{n-2}, \ldots, s, 1]'\end{aligned}$$

where $\boldsymbol{c}$ and $\boldsymbol{a}$ are the columns of c-coefficients and a-coefficients.

Moreover if

$$\mathrm{adj}(s\boldsymbol{I}-\boldsymbol{A}).\boldsymbol{Bq} = \boldsymbol{\Phi}_{n\times n}.[s^{n-1}, s^{n-2}, \ldots, s, 1]' \tag{6.4}$$

so that ϕ_{ij} is the coefficient of s^{n-j} in the ith element of the column $\mathrm{adj}(s\boldsymbol{I}-\boldsymbol{A}).\boldsymbol{Bq}$, then, on equating coefficients of like powers of s, we obtain

$$\boldsymbol{c}'-\boldsymbol{a}' = \boldsymbol{k}'\,\boldsymbol{\Phi} \quad \text{or} \quad \boldsymbol{k}' = (\boldsymbol{c}'-\boldsymbol{a}').\boldsymbol{\Phi}^{-1} \tag{6.5}$$

provided only that $\boldsymbol{\Phi}$ is nonsingular. It will now be shown that $\boldsymbol{\Phi}$ is always nonsingular provided that the pair $(\boldsymbol{A}, \boldsymbol{B})$ is completely controllable and provided that certain directions of the vector $\boldsymbol{q}$ are avoided.

For, using diagonal canonical form parameters

$$\begin{aligned}\mathrm{adj}(s\boldsymbol{I}-\boldsymbol{A}).\boldsymbol{Bq} &= (s\boldsymbol{I}-\boldsymbol{A})^{-1}\boldsymbol{Bq}.\det(s\boldsymbol{I}-\boldsymbol{A})\\ &= \boldsymbol{E}(s\boldsymbol{I}-\boldsymbol{\Lambda})^{-1}\boldsymbol{E}^{-1}\boldsymbol{Bq}.\det(s\boldsymbol{I}-\boldsymbol{\Lambda})\\ &= \boldsymbol{E}\,\mathrm{diag}\frac{\det(s\boldsymbol{I}-\boldsymbol{\Lambda})}{s-\lambda_i}.\boldsymbol{\beta q}\end{aligned}$$

Hence in eqn. 6.4, which defines $\boldsymbol{\Phi}$, we obtain

$$\mathrm{diag}\frac{\det(s\boldsymbol{I}-\boldsymbol{\Lambda})}{s-\lambda_i}.\boldsymbol{\beta q} \equiv \boldsymbol{E}^{-1}\boldsymbol{\Phi}[s^{n-1}, s^{n-2}, \ldots, s, 1]'$$

On the right hand side of this identity, if and only if $\boldsymbol{\Phi}$, and therefore also $\boldsymbol{E}^{-1}\boldsymbol{\Phi}$, is nonsingular will the elements of the resulting column matrix form a linearly independent set of polynomials in s. On the left hand side, the n quantities $\dfrac{\det(s\boldsymbol{I}-\boldsymbol{\Lambda})}{s-\lambda_i}$ are themselves an l.i. set of polynomials in s since $\det(s\boldsymbol{I}-\boldsymbol{\Lambda}) = (s-\lambda_1)(s-\lambda_2)\ldots(s-\lambda_n)$ and the eigenvalues are assumed distinct; moreover this l.i. property is maintained after each such quantity is multiplied by the corresponding element of $\boldsymbol{\beta q}$ *provided that no element of* $\boldsymbol{\beta q}$ *is zero.* Hence $\boldsymbol{\Phi}$ is nonsingular, if and only if no element of $\boldsymbol{\beta} \boldsymbol{q}$ is zero, i.e. if and only if (*a*) $\boldsymbol{\beta}$ has no null row, so that $(\boldsymbol{A}, \boldsymbol{B})$ is a completely controllable pair, and (*b*) $\boldsymbol{q}$ is not orthogonal to any row of $\boldsymbol{\beta}$, which is a very mild constraint on $\boldsymbol{q}$. Hence:

Theorem 6.2 Provided that the plant is completely controllable, all closed-loop poles may be arbitrarily assigned and the required feedback matrix is $(-\boldsymbol{H}) = -\boldsymbol{q}\boldsymbol{k}' = -\boldsymbol{q}(\boldsymbol{c}' - \boldsymbol{a}')\boldsymbol{\Phi}^{-1}$ where $\boldsymbol{q}$ is any vector not orthogonal to any row of $\boldsymbol{\beta}$ and $\boldsymbol{\Phi}$ is given by eqn. 6.4.

6.3.4.2 Some comments

(i) Finding $\boldsymbol{H}$ requires, in this case, the expansion of $\det(s\boldsymbol{I} - \boldsymbol{A})$ to give $\boldsymbol{a}'$, the required closed-loop polynomial to give $\boldsymbol{c}'$, the evaluation of $\operatorname{adj}(s\boldsymbol{I} - \boldsymbol{A})$ so that, with any given $\boldsymbol{q}$, $\boldsymbol{\Phi}$ may be obtained from eqn. 6.4, and preferably the evaluation of $\boldsymbol{\beta} = \boldsymbol{E}^{-1}\boldsymbol{B}$ so as to avoid banned directions of $\boldsymbol{q}$.

(ii) For a given plant and a given c.l.m.p., $\boldsymbol{\Phi}$ is a function of $\boldsymbol{q}$ only. The *scale* of $\boldsymbol{q}$, however, does not affect $\boldsymbol{H}$, for if $\boldsymbol{q}$ is multiplied by some constant, so is $\boldsymbol{\Phi}$: and since $\boldsymbol{H}$ contains the factors $\boldsymbol{q}$ and $\boldsymbol{\Phi}^{-1}$ it will be unchanged. Thus only the *direction* of $\boldsymbol{q}$ affects $\boldsymbol{H}$ and this direction is determined by the $(m - 1)$ mutual ratios of the elements of $\boldsymbol{q}$; in fact, of the $(n + m - 1)$ degrees of freedom in $\boldsymbol{H}$, n have been used to control the n coefficients of the c.l.m.p. Any selected $\boldsymbol{q}$ determines $\boldsymbol{\Phi}$ and $\boldsymbol{k}$ uniquely; hence an infinite number of $(\boldsymbol{q}, \boldsymbol{k})$ pairs exist all giving the same c.l.m.p., $\det(s\boldsymbol{I} - \boldsymbol{A} + \boldsymbol{B}\boldsymbol{H})$, though with different values of $\boldsymbol{H}$. The values of $\operatorname{adj}(s\boldsymbol{I} - \boldsymbol{A} + \boldsymbol{B}\boldsymbol{H})$ will however be different and the remaining $(m - 1)$ degrees of freedom inherent in the choice of $\boldsymbol{q}$ may therefore be used to modify the numerator of the closed-loop transfer matrix i.e. to influence other aspects of performance than stability. The precise effect of varying the direction of $\boldsymbol{q}$ upon these other aspects of performance can only be assessed by trial and error, which is probably very tedious, or by leaving $\boldsymbol{q}$ unknown and calculating $\boldsymbol{\Phi}$, $\boldsymbol{k}$ and hence $\boldsymbol{H}$ as explicit functions of $\boldsymbol{q}$, which is liable to be a very cumbersome algebraic process unless m and n have relatively low values.

6.3.4.3 An important extension

The $(m - 1)$ degrees of freedom inherent in the almost free choice of $\boldsymbol{q}$ may be added to by the following artifice, which is not previously mentioned. It will also serve the purpose of modifying the order of the external input $\boldsymbol{r}$, which, in all the theory so far presented in this Chapter, has necessarily been assumed equal to the order $(m \times 1)$ of the plant input $\boldsymbol{u}$ whereas it is normally equal to the order $(p \times 1)$ of the plant output $\boldsymbol{v}$.

Suppose then an external input vector $\boldsymbol{r}_{p\times 1}$ and assume that any given feedback matrix $(-\boldsymbol{H})$ is factorised as

$$-\boldsymbol{H}_{m\times n} = -\boldsymbol{M}_{m\times p}.\boldsymbol{N}_{p\times n}$$

Insert only $(-\boldsymbol{N})$ in the feedback path from the state $\boldsymbol{x}$, so that $(-\boldsymbol{Nx})$ is additive to the external input $\boldsymbol{r}$, and place $\boldsymbol{M}$ in the forward path between $\boldsymbol{r}$ and the plant input $\boldsymbol{u}$. We then have

$$\boldsymbol{u} = \boldsymbol{M}(\boldsymbol{r}-\boldsymbol{Nx})$$

and

$$\dot{\boldsymbol{x}} = \boldsymbol{Ax}+\boldsymbol{Bu} = (\boldsymbol{A}-\boldsymbol{BMN})\boldsymbol{x}+\boldsymbol{BMr} = (\boldsymbol{A}-\boldsymbol{BH})\boldsymbol{x}+\boldsymbol{BMr}.$$

The c.l.m.p. is therefore unchanged by the factorisation and is still $\det(s\boldsymbol{I}-\boldsymbol{A}+\boldsymbol{BH})$. But since $\boldsymbol{v}=\boldsymbol{Cx}$, the overall transfer matrix relating $\boldsymbol{V}(s)$ to $\boldsymbol{R}(s)$ is now $\boldsymbol{C}.\text{adj}(s\boldsymbol{I}-\boldsymbol{A}+\boldsymbol{BH}).\boldsymbol{BM}/\det(s\boldsymbol{I}-\boldsymbol{A}+\boldsymbol{BH})$ and is modified by the addition of the factor $\boldsymbol{M}$ in the numerator. Since the factorisation of $\boldsymbol{H}$ may in general be carried out in a large number of ways, this additional factor provides further degrees of freedom in the design.

6.4 Output feedback

As might be expected, if the feedback originates from the output $\boldsymbol{v}_{p\times 1}$ instead of from the state $\boldsymbol{x}_{n\times 1}$, then if, as is normal, $p<n$, we shall not be able to secure such a firm control over the closed-loop poles. A further paper by Fallside and Seraji[5] treats this problem and we shall outline its essentials.

Representing the plant as before by

$$\dot{\boldsymbol{x}} = \boldsymbol{Ax}+\boldsymbol{Bu}, \boldsymbol{v} = \boldsymbol{Cx},$$

with rank $\boldsymbol{B}=m$, rank $\boldsymbol{C}=p, m\leqslant n$ and $p<n$, we visualise a feedback matrix $(-\boldsymbol{H})_{m\times p}$ from $\boldsymbol{v}$ to $\boldsymbol{u}$, with, as before, an external input $\boldsymbol{r}_{m\times 1}$ also contributing to $\boldsymbol{u}$. Then

$$\dot{\boldsymbol{x}} = \boldsymbol{Ax}+\boldsymbol{B}(\boldsymbol{r}-\boldsymbol{Hv}) = (\boldsymbol{A}-\boldsymbol{BHC})\boldsymbol{x}+\boldsymbol{Br}$$

so that the c.l.m.p. is $\det(s\boldsymbol{I}-\boldsymbol{A}+\boldsymbol{BHC})$ and the overall transfer matrix relating $\boldsymbol{V}(s)$ to $\boldsymbol{R}(s)$ is

$$\boldsymbol{F}(s) = \boldsymbol{C}.\text{adj}(s\boldsymbol{I}-\boldsymbol{A}+\boldsymbol{BHC}).\boldsymbol{B}/\det(s\boldsymbol{I}-\boldsymbol{A}+\boldsymbol{BHC}).$$

The rank of $\boldsymbol{H}$ is again assumed to be unity so that we may write

$$\boldsymbol{H}_{m\times p} = \boldsymbol{q}_{m\times 1}.\boldsymbol{k}'_{1\times p}$$

Following the same line of analysis as in Section 6.3.4.1, we have first

$$\text{c.l.m.p.} = \det(sI - A + BHC) = \det(sI - A) + k'C.\text{adj}(sI - A).Bq$$

(see Appendix)

Write

$$C.\text{adj}(sI - A).Bq \equiv \Psi_{p \times n}[s^{n-1}, s^{n-2}, \ldots, s, 1]'$$

so that Ψ_{ij} is the coefficient of s^{n-j} in the ith element of the column $C.\text{adj}(sI - A).Bq$. There results

$$k'\Psi[s^{n-1}, s^{n-2}, \ldots, s, 1]' \equiv (c' - a').[s^{n-1}, s^{n-2}, \ldots, s, 1]'$$

so that

$$k'\Psi \cdot = c' - a'$$

or

$$\Psi'k = c - a \equiv \delta \text{, say.} \tag{6.6}$$

Note in passing that $\Psi = C\Phi$; it follows that if Φ is nonsingular then Ψ is of maximal rank p, since C is. The nonsingularity of Φ is, however, not a necessary condition, though it is sufficient. Whatever the reasons, we shall suppose that rank $\Psi = p$.

If we pre-assign certain real values to the elements of q, then Ψ is uniquely calculable and eqn. 6.6 then represents a set of n linear equations for the p elements of k, since c and a and therefore δ are assumed known. Since $n > p$, these equations will only be soluble if they are consistent. Since $(\Psi')_{n \times p}$ is of rank p, it contains at least one set of p linearly independent rows. Hence, re-arranging the n scalar equations of eqn. 6.6 if necessary so that the *first p* rows of Ψ' are l.i., we may rewrite eqn. 6.6 in the partitioned form

$$\Psi'k \equiv \begin{bmatrix} \theta_1 \\ \theta_2 \end{bmatrix} . k = \begin{bmatrix} \delta_1 \\ \delta_2 \end{bmatrix}$$

where

θ_1 is square, of order p and nonsingular
θ_2 is of order $(n-p) \times p$, δ_1 of order $p \times 1$
δ_2 of order $(n-p) \times 1$

From the upper part of this equation we deduce the unique solution $k = \theta_1^{-1}.\delta_1$ which, if the equations are to be consistent, must also satisfy the remaining $(n-p)$ scalar equations. Hence the consistency condition is

$$\theta_2 . \theta_1^{-1} . \delta_1 = \delta_2 \tag{6.7}$$

which is equivalent to saying that every element of δ_2 is some linear combination of the elements of δ_1, i.e. using eqn. 6.6, some $(n-p)$ elements of c are linear functions of the other p elements. Hence only p of these elements may be assigned arbitrarily, i.e. only p of the closed-loop poles may be assigned arbitrarily. (Assigning a complex pole of course counts as assigning two poles, the assigned pole and its conjugate.) It may be, of course, that one or more of the $(n - p)$ unassigned poles will fall in undesirable positions, possibly even in the right-half-plane of s: if so we have to start again, either with a new set of assigned poles or with a new value of $\boldsymbol{q}$. Algorithms have been devised such that, if the n polar values required conflict with the consistency conditions of eqn. 6.7, a set of polar values *in the neighbourhood* of the required values may be found which satisfies these conditions, but often the 'neighbourhood' is not a very restricted area!

6.4.1 The influence of q

Since $\boldsymbol{H}$ is here of order $(m \times p)$ and rank unity, it has $(m + p - 1)$ degrees of freedom of which p have been used to fix p of the n closed-loop poles. As in the case of state feedback, we are left with $(m - 1)$ degrees of freedom, implicit in the direction of $\boldsymbol{q}$. Is it possible to use these to allocate further closed-loop poles?

Leave $\boldsymbol{q}$ unspecified; then every element in $\boldsymbol{C}.\text{adj}(s\boldsymbol{I} - \boldsymbol{A}).\boldsymbol{Bq}$ is clearly a linear combination of the elements of $\boldsymbol{q}$. The same is true therefore of the elements of $\boldsymbol{\Psi}$ and therefore of the elements of $\boldsymbol{\theta}_1$ and $\boldsymbol{\theta}_2$. (The elements of the left hand side column matrix of eqn. 6.6 are therefore linear combinations of the products $q_i k_j (i = 1, 2, \ldots m;$ $j = 1, \ldots p)$, showing that a scalar factor interchange between $\boldsymbol{q}$ and $\boldsymbol{k}$ does not affect the problem, as is obvious from $\boldsymbol{H} = \boldsymbol{qk}'$.) In eqn. 6.7, writing $\boldsymbol{\theta}_2 \boldsymbol{\theta}_1^{-1}$ as $\boldsymbol{\theta}_2 \text{ adj } \boldsymbol{\theta}_1 / \det \boldsymbol{\theta}_1$, the denominator will be homogeneous in the elements of $\boldsymbol{q}$, of degree p; the numerator is the product of $\boldsymbol{\theta}_2$ of which the elements are linear and homogeneous in the q_i, and adj $\boldsymbol{\theta}_1$, of which the elements are homogeneous in the q_i of degree $(p - 1)$: hance the elements of the numerator, like those of the denominator, are homogeneous of degree p, showing that the scale of $\boldsymbol{q}$ is irrelevant and only the direction of $\boldsymbol{q}$ is important.

Since changing this direction will modify the coefficient of $\boldsymbol{\delta}_1$ in eqn. 6.7, it might be hoped that, if this direction is suitably chosen, eqn. 6.7 might be satisfied for at any rate $(m - 1)$ assigned values of $\boldsymbol{\delta}_2$ i.e. for a total of $(p + m - 1)$ values of $\boldsymbol{\delta}$ and therefore of $\boldsymbol{c}$. If this were the case, then if $p + m - 1 > n$, we could assign all the closed-loop poles (and possibly have some degrees of freedom left!). This hope may be realised in some cases but in others is not. The difficulty lies mainly in the fact that the q_i do not appear linearly in eqn. 6.7 and that, from the point of view of physical realisability, we are tied to real values of $\boldsymbol{H}$ and therefore of $\boldsymbol{q}$ and $\boldsymbol{k}$. Nonlinear equations unfortunately do not always have real roots, even when, as here, they have real coefficients. As a simple example, with $m = p = 2$ and $n = 3$, suppose that eqn. 6.6 takes the simple form

$$\begin{bmatrix} 0 & q_1 \\ q_1 & q_2 \\ q_2 & 0 \end{bmatrix} \begin{bmatrix} k_1 \\ k_2 \end{bmatrix} = \begin{bmatrix} \delta_1 \\ \delta_2 \\ \delta_3 \end{bmatrix}$$

The first equation gives $k_2 = \delta_1/q_1$, the third gives $k_1 = \delta_3/q_2$: to satisfy the second therefore requires $(q_1/q_2)^2\delta_3 - (q_1/q_2)\delta_2 + \delta_1 = 0$. If q_1/q_2 is to be real, this requires $\delta_2^2 \geqslant 4\delta_1\delta_3$: this additional constraint on the elements of $\boldsymbol{\delta}$ and hence on the elements of $\boldsymbol{c}$ may, or may not, be satisfied by the proposed closed-loop poles.

Nevertheless the method of unassigned $\boldsymbol{q}$ is probably worth trying, though, as in the case of state feedback, the algebra becomes very complex unless m, p and n have low values. The factorisation method of Section 6.3.4.3 (in this case $\boldsymbol{H}_{m\times p} = \boldsymbol{M}_{m\times p} \cdot \boldsymbol{N}_{p\times p}$) may of course be used, but only modifies the numerator of the transfer matrix and has no effect upon the poles. To assign more poles may require the additional complication of Luenberger observers which, subject to certain conditions, restore the problem to one of state feedback.

If on the other hand we increase the number of degrees of freedom by dropping the unity-rank constraint on H, $\det(s\boldsymbol{I} - \boldsymbol{A} + \boldsymbol{BH})$ is no longer linear in the elements of $\boldsymbol{H}$ and we have to find real solutions to a number of simultaneous nonlinear equations; moreover if $n > mp$, there are still consistency conditions to be satisfied, and in any case there is no guarantee that as many as mp poles may be arbitrarily assigned.

6.5 Luenberger observers[6,7]

A Luenberger observer is, in practice, a dynamic simulator which has the property that, when fed with the plant input, $\boldsymbol{u}(t)$, and the plant output, $\boldsymbol{v}(t)$, as separate inputs, it generates a state-vector $\boldsymbol{z}(t)$ which, in the more general case, is some numeric matrix transformation $\boldsymbol{T}.\boldsymbol{x}(t)$ of the plant state-vector $\boldsymbol{x}(t)$. Since the observer is a simulator, it may be assumed that every element of its state-vector $\boldsymbol{z}(t)$ is accessible for feedback purposes. It will be seen in the sequel that $\boldsymbol{z}(t)$ is *normally an approximation* to $\boldsymbol{T}.\boldsymbol{x}(t)$.

6.5.1 Preliminary analysis

Suppose the plant to be representable by the standard equations:

$$\dot{x}_{n\times 1} = Ax + Bu_{m\times 1}; v_{p\times 1} = Cx + Du \qquad (6.8)$$

Let us postulate an observer dynamic equation, with z as state-vector and u and v as separate inputs, of the form

$$\dot{z}_{q\times 1} = Lz + Mu + Nv \qquad (6.9)$$

Then with T of order $q \times n$ it is required to approximate to $T.x(t)$ by $z(t)$.

Substituting for v in eqn. 6.9 from eqn. 6.8, we deduce

$$\dot{z} - T\dot{x} = Lz + Mu + NCx + NDu - TAx - TBu$$
$$= L(z - Tx) + (NC - TA + LT)x + (M + ND - TB)u$$

Hence if we can find L, M, N, to satisfy

$$TA - LT = NC, M = TB - ND, \qquad (6.10)$$

the previous equation reduces to $\dot{z} - T\dot{x} = L(z - Tx)$ which has the solution

$$z(t) - T.x(t) = \exp(Lt).\{z(0) - T.x(0)\} \qquad (6.11)$$

Hence if eqn. 6.10 can be satisfied in such a way that all eigenvalues of L have negative real parts, the difference between $z(t)$ and $T.x(t)$ will merely consist of the transient terms on the right hand side of eqn. 6.11 and, onee these have died away, $z(t)$ will equal $T.x(t)$. To this extent $z(t)$ is an *approximation* to $T.x(t)$.

Note that the first of the conditions for eqn. 6.10 is independent of B and D and must be satisfied even when the system is 'free', i.e. $u = 0$. Considered as an equation in T (qn scalar equations for the qn elements of T) it may be shown that T may be found uniquely provided only that A and L have no common eigenvalue; thus, choosing N arbitrarily and selecting L, with this minor constraint, to have eigenvalues with sufficiently negative real parts, T may be found uniquely; the second part of eqn. 6.10 then gives M, N being arbitrary throughout.

6.5.2 The 'identity' observer

Producing an approximation to $T.x(t)$ can only lead to $x(t)$ itself if$T$ has an inverse, i.e. if T is square so that $q = n$. Rather than produce $T.x(t)$ and then pass it through a subsystem of transfer matrix T^{-1}, a more direct approach is clearly to make $q = n$ and $\boldsymbol{T} = \boldsymbol{I}_n$ in the above theory. We then require

$$\boldsymbol{L} = \boldsymbol{A} - \boldsymbol{N}\boldsymbol{C}, \ \boldsymbol{M} = \boldsymbol{B} - \boldsymbol{N}\boldsymbol{D} \tag{6.12}$$

and, provided that these are satisfied,

$$\boldsymbol{z}(t) - \boldsymbol{x}(t) = \exp(\boldsymbol{L}t).\{\boldsymbol{z}(0) - \boldsymbol{x}(0)\}.$$

It may be shown that in the first part of eqn. 6.12, *the eigenvalues* $\boldsymbol{L}$ *may be chosen arbitrarily* (provided of course, since $\boldsymbol{L}$ is real, that complex eigenvalues occur only as conjugate complex pairs) *if and only if the matrix pair* (C, A) *are completely observable*, i.e. rank $[C', A'C', \ldots, (A')^{n-1}C'] = n$. Thus if the plant is completely observable, the rate of decay of the transient terms in eqn. 6.11, with $\boldsymbol{T} = \boldsymbol{I}$, may be made as rapid as we please. On the other hand, as Luenberger points out, making this rate of decay too rapid, i.e. making the eigenvalues of $\boldsymbol{L}$ too negative in their real parts, leads to the observer acting as an approximate differentiator so that high-frequency noise is over-amplified: Luenberger suggests eigenvalues of $\boldsymbol{L}$ only slightly more negative than other eigenvalues present in the observed system. [Even if $(\boldsymbol{C}, \boldsymbol{A})$ is not a completely observable pair, it may be that satisfactory values for the eigenvalues of $\boldsymbol{L}$ may be found by a suitable choice of the arbitrary N].

6.5.3 Reduced dimension observer

The identity observer just considered is unnecessarily complex (and expensive) to the extent that since some of the elements of $\boldsymbol{x}(t)$ are already available in the plant output $\boldsymbol{v}(t)$, it appears unnecessary to reproduce *all* these elements in the state-vector $\boldsymbol{z}(t)$ of the observer. Suppose for simplicity, which is often the case in the original state-representation of a system, that the output elements of the plant are some p of the n elements of its state-vector $\boldsymbol{x}$. Then by re-ordering the elements of $\boldsymbol{x}$ if necessary, we may write $\boldsymbol{x}$ in the partitioned form $\boldsymbol{x} = \begin{bmatrix} \boldsymbol{v} \\ \boldsymbol{w} \end{bmatrix}$ where $\boldsymbol{v}$, as before, is the output vector and $\boldsymbol{w}$, of order $(n-p) \times 1$, consists of those elements of $\boldsymbol{x}$ not appearing in the output. Partitioning $\boldsymbol{A}$ and $\boldsymbol{B}$, the plant dynamic equation becomes

$$\begin{bmatrix} \dot{\boldsymbol{v}} \\ \dot{\boldsymbol{w}} \end{bmatrix} = \begin{bmatrix} \boldsymbol{A}_{11} & \boldsymbol{A}_{12} \\ \boldsymbol{A}_{21} & \boldsymbol{A}_{22} \end{bmatrix} \begin{matrix} \boldsymbol{v} \\ \boldsymbol{w} \end{matrix} + \begin{bmatrix} \boldsymbol{B}_1 \\ \boldsymbol{B}_2 \end{bmatrix} \boldsymbol{u},$$

while its output equation becomes

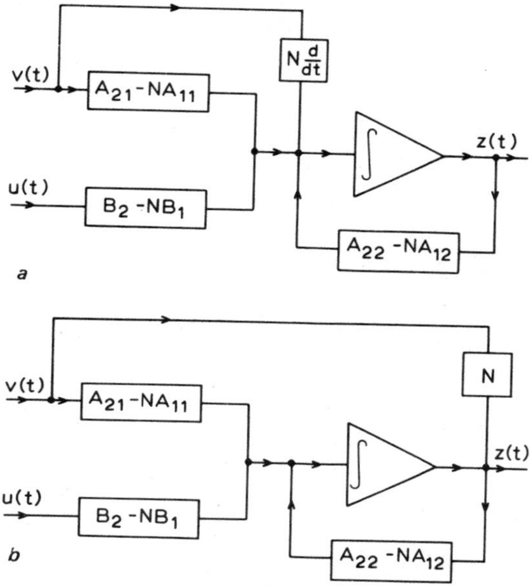

Fig. 6.2 Schematic of observer
(*a*) including a differentiation
(*b*) without the differentiation

$$v = [I_p \quad 0]\begin{bmatrix} v \\ w \end{bmatrix}$$

Breaking up the dynamic equation into two separate equations, we deduce:

$$\begin{aligned} \dot{w} &= A_{22}w + (A_{21}v + B_2u) \\ A_{12}w &= \dot{v} - A_{11}v - B_1u \end{aligned} \tag{6.13}$$

Since v, and therefore $\dot{v}$, is known, as also is u, $A_{12}w$ may be considered as known for any given u. We shall therefore consider eqn. 6.13 as the state-equations of a system for which the state-vector is w, A is replaced by A_{22}, the input is $(A_{21}v + B_2u)$ so that B is replaced by I_{n-p}, and the output is $A_{12}w$, so that C is replaced by A_{12} and $D = 0$, this output, incidentally, being given by the second expression of eqn. 6.13. We wish to design an identity observer for this system. Using the result given in eqn. 6.12 we deduce

$$\dot{z} = (A_{22} - NA_{12})z + (I_{n-p} - NO)(A_{21}v + B_2u) + NA_{12}\dot{w}$$

or, substituting for w from the second part of eqn. 6.13:

$$\dot{z} = (A_{22} - NA_{12})z + A_{21}v + B_2 u + N(\dot{v} - A_{11}v - B_1 u) \quad (6.14)$$

It may be shown that if (C, A) is completely controllable, so is (A_{12}, A_{22}). If then the plant is completely controllable, the eigenvalues of $(A_{22} - NA_{12})$ may be fixed arbitrarily by a suitable choice of the unknown matrix N. The observer represented by eqn. 6.14 may be simulated as in Fig. 6.2*a*. It has the disadvantage that the simulation requires a differentiation operation.

This operation may however be removed by the rather obvious change to Fig. 6.2*b*. Both diagrams correspond to eqn. 6.14. As a check, subtracting the first expression of eqn. 6.13 leads to

$$\begin{aligned} \dot{z} - \dot{w} &= (A_{22} - NA_{12})(z - w) - NA_{12}w - NA_{11}v - NB_1 u + N\dot{v} \\ &= (A_{22} - NA_{12})(z - w) \end{aligned}$$

the remaining terms summing to zero by virtue of the second part of eqn. 6.13. Thus the observer state-vector approximates, as time increases, to w, which is the nonaccessible part of the plant state-vector. Any desired feedback from the plant state-vector, say $(-H).x = -[H_1 \quad H_2]\begin{bmatrix} v \\ w \end{bmatrix} = -(H_1 v + H_2 w)$ may thus be approximated to by $-(H_1 v + H_2 z)$, a feedback taken jointly from the plant output and the observer output.

6.5.4 Effect of observer on closed-loop modes

It is of course important to verify that the use of a dynamic observer does not adversely affect the stability of the closed-loop system. It may, in fact, be shown that the closed-loop modes are a combination of the modes that would be present if an observer could be dispensed with and the modes of the observer itself.

Considering for instance the reduced order observer of the last Section, suppose we take a feedback $-(H_1 v + H_2 z)$ to the input of the plant so that $u = -(H_1 v + H_2 z)$. Then from eqn. 6.13 we derive

$$\begin{aligned} \dot{v} &= (A_{11} - B_1 H_1)v + A_{12}w - B_1 H_2 z \\ &= (A_{11} - B_1 H_1)v + (A_{12} - B_1 H_2)w - B_1 H_2(z - w) \\ \dot{w} &= (A_{21} - B_2 H_1)v + A_{22}w - B_2 H_2 z \\ &= (A_{21} - B_2 H_1)v + (A_{22} - B_2 H_2)w - B_2 H_2(z - w) \end{aligned}$$

while it was shown above that

Chapter 7

The commutative controller and dyadic transfer matrices

7.1 The commutative controller

This method of controller design was suggested by MacFarlane in 1970.[1] It suffers from a number of weaknesses[2,3] but if used with care, it can give useful results. Like many other controller design techniques, the method seeks to reduce the problem to the design of a number of noninteracting control loops, each of which may then be designed using the familiar classical methods for one-input/one-output systems.

It is supposed in the first place that $m = p$ so that, in Fig. 5.1, $\boldsymbol{P}(s)$, $\boldsymbol{K}(s)$ and $\boldsymbol{H}(s)$ are all square of order p. It is supposed further that the eigenvalues of $\boldsymbol{P}(s)$ are either distinct or, if not, that any multiple eigenvalue is associated with as many l.i. eigenvectors as its multiplicity: in either case we may write

$$\boldsymbol{P}(s) = \boldsymbol{E}(s).\mathrm{diag}\, p_i(s).\boldsymbol{E}^{-1}(s) \tag{7.1}$$

where the $p_i(s)$ are the eigenvalues and $\boldsymbol{E}(s)$ is the eigenvector assembly matrix. The key to the method is to suppose that $\boldsymbol{K}(s)$ and $\boldsymbol{H}(s)$ are constrained to have the same eigenvector matrix as $\boldsymbol{P}(s)$ but associated with different eigenvalues; in other words we suppose

$$\boldsymbol{K}(s) = \boldsymbol{E}(s).\mathrm{diag}\, k_i(s).\boldsymbol{E}^{-1}(s) \quad \text{and} \quad \boldsymbol{H}(s) = \boldsymbol{E}(s), \mathrm{diag}\, h_i(s).\boldsymbol{E}^{-1}(s) \tag{7.2}$$

The constraint is obvious: for $\boldsymbol{K}$ and $\boldsymbol{H}$ are now determined by m eigenvalues instead of by m^2 elements.

It is clear that any matrices of this type commute under multiplication (hence the name of the method). For instance, dropping the functional (s) for simplicity, $\boldsymbol{KH} = \boldsymbol{E}.\mathrm{diag}\, k_i h_i.\boldsymbol{E}^{-1} = \boldsymbol{E}.\mathrm{diag}\, h_i k_i.\boldsymbol{E}^{-1} = \boldsymbol{HK}$. Indeed it is easy to show that if $\phi(\boldsymbol{P}, \boldsymbol{K}, \boldsymbol{H})$ is any algebraic function of $\boldsymbol{P}$, $\boldsymbol{K}$ and $\boldsymbol{H}$ then $\phi(\boldsymbol{P}, \boldsymbol{K}, \boldsymbol{H}) = \boldsymbol{E}.\mathrm{diag}\, \phi(p_i, k_i, h_i).\boldsymbol{E}^{-1}$. In particular, the closed-loop transfer matrix $F(s)$ is given by

$$F = (I + PKH)^{-1}PK = E.\text{diag}\,\frac{p_i k_i}{1 + p_i k_i h_i}.E^{-1} \equiv E.\text{diag}\,f_i.E^{-1} \tag{7.3}$$

where

$$f_i = \frac{p_i k_i}{1 + p_i k_i h_i} \tag{7.4}$$

Moreover, since $V(s) = F(s)R(s)$,

$$(E^{-1}V) = E^{-1}FR = \text{diag}\,f_i.(E^{-1}R)$$

showing that the *transformed* output, $E^{-1}V$, is related to the *transformed* external input, $E^{-1}R$, by a diagonal, and therefore noninteracting, transfer matrix. The method of design is to choose each $k_i(s)$ and its associated $h_i(s)$ so that each $f_i(s)$ is a satisfactory closed-loop transfer function, this selection process being carried out by classical methods (Nyquist loci, gain margins, phase margins, root-locus methods etc.)

7.2 Comments

(i) It is not very clear why satisfactory $f_i(s)$ relationships between corresponding elements of the *transformed* output and input vectors should necessarily lead to satisfactory relationships (*not* of a noninteracting nature) between the *actual* output and input vectors. Note however that det $F(s) = \prod_{i=1}^{p} f_i(s)$, that the poles of $F(s)$, which determine closed-loop stability, have the same location as those of det $F(s)$ and that these are, in turn, the poles of the several $f_i(s)$. Since, with a satisfactory design, each $f_i(s)$ will have its poles in the left half-plane of s, it follows that the closed-loop system will be stable. But there is no guarantee at all that the degree of interaction in $F(s)$ will be reasonable, even though the *transformed* equivalent of $F(s)$, namely diag $f_i(s)$, is of course free of interaction. It is largely this high degree of interaction which doomed the solution to MacFarlane's original example.

(ii) The second important criticism is that although the elements of $P(s)$ are normally rational functions of s, this is very often not true of its eigenvalues. If any eigenvalues contain irrational functions of s, the eigenvector assembly matrix $E(s)$ will also contain such elements. Then, in designing $K(s)$ for instance, either we shall be forced to choose some irrational $k_i(s)$ in order to make $K(s)$ rational (which will greatly complicate the design of the associated ith loops), or we make the k_i rational, in which case K will contain irrational elements and will be difficult to realise as a physical unit. On the other hand, to limit the method to plant matrices with rational eigenvalues is very restrictive.

7.2 On dyadic matrices

The theory that follows is applicable whatever the field over which the matrices exist: real numbers, complex numbers, rational functions of s etc. Although we shall later be mainly concerned with the field of rational functions of s, for the present we shall omit the functional symbol (s) and leave the field unspecified. We define:

A dyad D of order $m \times n$ is a matrix of this order which can be expressed as a (column).(row) product, $D_{m\times n} = y_{m\times 1} z'_{1\times n}$.

The rank of a dyad can therefore not exceed unity but, if y or z is null, may be zero, in which case D is null. It is easy to prove moreover that any matrix of rank unity (or zero) can be expressed as a column-row product and is therefore a dyad.

It is clearly always possible to extract some nonzero scalar factors k_1 from y and k_2 from z' and to write, with $k = k_1 k_2, D = k.y_1.z'_1$, where $y_1 = y/k_1, z'_1 = z'/k_2$. Thus, more generally, *a dyad may be defined as the product of a scalar, a column and a row*. This apparently trivial extension is of importance in what follows.

7.2.1 Dyadic expansion of matrices

Theorem 7.1 The triple product $M_{m\times n} = Y_{m\times p}.(\text{diag } m_i)_{p\times p}.Z_{p\times n}$ is equal to $\sum_{i=1}^{p} m_i y_i z'_i$, where y_i is the ith column of Y and z'_i is the ith row of Z (i.e. the ith column of Z').

Proof: The product $Y.\text{diag } m_i$ multiplies the ith column of Y by m_i so that the coefficient of a particular m_i is a matrix which is null except for its ith column which is y_i. When this matrix is postmultiplied by Z, the result is therefore a matrix of which the jth row is $y_{ji}.z'_i$, namely the dyad $y_i.z'_i$. This dyad is therefore the coefficient of m_i in the given triple product, which proves the theorem.

Comments

(i) The theorem expresses the triple product as the sum of p dyads of order $m \times n$. It is interesting to note that if we make every m_i equal to unity, so that diag $m_i = I_p$, we obtain

$$Y.Z = \sum_{i=1}^{p} y_i z_i'$$

which expresses the product of two matrices in terms of *columns* of the first factor and *rows* of the second. (The usual definition of a product gives the *elements* of the product in terms of the rows of the first factor and the columns of the second.)
(ii) If $m = n = p$, then M, Y, Z and the dyads in the summation are all square of order p; this is the particular case which will mainly concern us.

7.2.1.1 Minimal dyadic expansions

Consider now *any* square matrix M of order p and rank $R \leqslant p$. Take *any* nonsingular matrix A of the same order. Then $M = A.A^{-1}M = AA_1$ where $A_1 = A^{-1}M$. Since A is nonsingular, rank $A_1 = R$, i.e. some R of the p rows of A_1 form a linearly independent (l.i.) set of which the remaining $(p - R)$ rows are linear combinations. For convenience, rearrange the rows of A_1 if necessary so that *the first* R rows form an l.i. set: this may be done through the transformation $A_2 = CA_1$, where C is a so-called permutation matrix, which is null except that every row and every column contains one and only one element of value unity. Then the last $(p - R)$ rows of A_2 are linear combinations of the first R rows.

Now write $A_3 = KA_2$ where the matrix K represents a sequence of 'elementary row operations', i.e. the subtraction of a scalar multiple of one row of A_2 from another. Choose K in this case so that the first R rows of A_2 are left unchanged but so that from any of the remaining $(p - R)$ rows of A_2 is subtracted precisely that linear combination of the first R rows of A_2 to which it is equal. Then the first R rows of A_3 and A_2 are identical but the last $(p - R)$ rows of A_3 are null: hence we may write $A_3 = I_{p,R}.A_4$, where $I_{p,R}$ denotes a depleted identity matrix of order p in which only the first R diagonal elements are unity, the rest zero. The first R rows of A_3 and A_4 are therefore identical and the last $(p - R)$ rows of A_4 are arbitrary, since their elements are all multiplied by zero in forming A_3: choose these rows in any manner which makes A_4 nonsingular. It is easily shown that det $C = \pm 1$ and det $K = 1$, so that C and K are nonsingular. Then

$$M = AA_1 = AC^{-1}A_2 = AC^{-1}K^{-1}A_3 = AC^{-1}K^{-1}I_{p,R}.A_4 \equiv Y.I_{p,R}.Z \tag{7.5}$$

where $Y = AC^{-1}K^{-1}$ and $Z = A_4$ so that both are nonsingular. Hence M, which is of order p and rank R, is always *equivalent* to $I_{p,R}$, that is, may be obtained from it by pre- and postmultiplication by nonsingular matrices. This is a well-known theorem in matrix theory.

Note that since A is completely arbitrary, apart from being nonsingular, and clearly determines Y, and since in addition the last $(p-R)$ rows of Z are virtually arbitrary, Y and Z may be found in an infinite number of ways. Moreover, applying Theorem 7.1 to eqn. 7.5 gives

$$M = Y.I_{p,R}.Z = \sum_{i=1}^{R} y_i.z_i' \tag{7.6}$$

which shows that any matrix of order p and rank R may be expanded as the sum of R dyads of the same order, in which, since Y and Z are nonsingular, the set of columns, y_i, and the set of rows, z_i', $(i = 1, 2, \ldots, R)$ are both l.i. sets. The number of dyads can never be less than R since, whatever values of Y and Z are taken, no y_i and no z_i can therefore be null. We define:

A minimal dyadic expansion (MDE) of a square matrix M *of rank* R *is any sum of* R *dyads such that their column factors and their row factors are both l.i. sets.*

The number of terms in a dyadic expansion of M can always be made greater than R by, for instance, expressing any column (or row) factor as the sum of two or more columns (or rows), but the resulting set of columns (or rows) will no longer be l.i. and the expansion will therefore not be minimal.

Finally, it is important to note that, using the scalar factor extraction method of Section 7.2, the MDE in eqn. 7.6 may be generalised to

$$M = \sum_{i=1}^{R} m_i y_i z_i' = Y.\text{diag}_R m_i.Z \tag{7.7}$$

where Y, Z are non-singular, no m_i is zero and $\text{diag}_R m_i$ is a diagonal matrix of order p in which the last $(p-R)$ diagonal elements are zero. The y_i and z_i' in eqn. 7.7 are of course nonzero scalar multipliers of the y_i and z_i' in eqn. 7.6.

7.2.2 On dyadic transfer matrices (DTM's)

The use of dyadic expansions in controller design and, in particular, the concept of the dyadic transfer matrix, appear to be due to Owens.[4] Although, for physical realisability, $P(s)$, $K(s)$ and $H(s)$ must be matrices over the field of *real* rational functions of s, we prefer at first to consider matrices over the field of complex functions of s, which, of course, contains the field of real rational functions as a subfield. For ease of presentation, the analysis will be confined to nonsingular matrices, i.e. $R = p$, though many of the results that follow may be extended to the case $R < p$.

If then $M(s)$ is a square, nonsingular matrix of order p over the field of complex rational functions of s, $M(s)$ can be expressed, in general in an infinite number of ways, as

$$M(s) = Y(s).\text{diag}\, m_i(s).Z(s) \qquad (i = 1, 2, \ldots, p) \tag{7.8}$$

where $Y(s), Z(s)$ are nonsingular and where no $m_i(s) = 0$. We define:

A dyadic transfer matrix (DTM) is a square, nonsingular matrix of any order p, over the field of real rational functions of s (and therefore over part of the broader field of complex rational functions) capable of expression, in terms of matrices over this broader field, as

$$M(s) = Y.\text{diag}\, m_i(s).Z \tag{7.9}$$

where Y, Z *are numeric (field of complex numbers) and nonsingular.*

Note that although the definition is only concerned with *real* $M(s)$, any or all of its three factors are allowed to be complex. But the main restriction on $M(s)$, which eliminates many transfer matrices, is that Y and Z must be numeric, the only factor which is a function of s being diag $m_i(s)$. This restriction leads to the idea of approximating to a matrix which is not a DTM by one which is, a problem briefly discussed later.

Owens adds to the definition the further restriction that $M^{-1}(0)$ shall exist. *This is an unnecessary restriction and is omitted here.*

7.2.2.1 *Interesting properties of DTM's*

Theorem 7.2 If $M(s)$ *is a DTM and if* c *is any real number such that* $M(c)$ *is finite and nonsingular, then the quantities* $m_i(s)/m_i(c)$ *are the eigenvalues of both* $M(s)M^{-1}(c)$ *and* $M^{-1}(c)M(s)$*. Moreover the columns of* Y *are associated eigenvectors of* $M(s)M^{-1}(c)$ *and the rows of* Z *are associated pre-multiplying eigenvectors of* $M^{-1}(c)M(s)$.

(If A is a matrix with eigenvalue λ then if e is such that $Ae = \lambda e$, e is an eigenvector; if f' is such that $f'A = \lambda f'$, then f' is a premultiplying eigenvector of A and is clearly an eigenvector of A' since $A'f = \lambda f$.)

Proof: From eqn. 7.9,

$$\begin{aligned} M(s)M^{-1}(c) &= Y.\text{diag}\, m_i(s).Z.Z^{-1}.\text{diag}\,\{1/m_i(c)\}.Y^{-1} \\ &= Y.\text{diag}\,\{m_i(s)/m_i(c)\}.Y^{-1} \\ M^{-1}(c)M(s) &= Z^{-1}.\text{diag}\,\{1/m_i(c)\}.Y^{-1}.Y\,\text{diag}\, m_i(s).Z \\ &= Z^{-1}.\text{diag}\,\{m_i(s)/m_i(c)\}.Z \end{aligned}$$

Both these products are therefore obtained from diag $\{m_i(s)/m_i(c)\}$ by a similarity transformation and therefore have the same eigenvalues, namely $m_i(s)/m_i(c), i = 1, 2, \ldots, p$. Moreover, the two equations may be written

$$\begin{aligned} M(s)M^{-1}(c).\,Y &= Y\,\text{diag}\,\{m_i(s)/m_i(c)\} \\ Z.M^{-1}(c)M(s) &= \text{diag}\,\{m_i(s)/m_i(c)\}.Z \end{aligned}$$

and equating the ith columns in the first and the ith rows in the second:

$$\begin{aligned} M(s)M^{-1}(c).y_i &= \{m_i(s)/m_i(c)\}.y_i \\ z_i'.M^{-1}(c)M(s) &= \{m_i(s)/m_i(c)\}.z_i' \end{aligned}$$

which completes the proof.

Note that if any of the eigenvalues are multiple, they must be associated with a corresponding number of l.i. eigenvectors, i.e. $M(s)M^{-1}(c)$ and $M^{-1}(c)M(s)$ must be diagonalisable by a similarity transformation. The theorem is a necessary property of a DTM but does not define a DTM.

Theorem 7.3 If $M(s)$ is a DTM then so are $(a)\, M^{-1}(s)$; $(b)\, M'(s)$; (c) $U.M(s).V$ where U, V are real, nonsingular and numeric; $(d)\, dM(s)/ds$, provided this matrix is nonsingular.

Proof: From eqn. 7.9

(a) $\quad M^{-1}(s) = Z^{-1}.\text{diag}\,\{1/m_i(s)\}\, Y^{-1}$; (b) $\; M'(s) = Z'.\text{diag}\, m_i(s).Y'$

(c) $U.M(s).V = UY.\text{diag}\, m_i(s).ZV$; $\qquad (d)\, \dfrac{dM(s)}{ds} = Y.\text{diag}\, \dfrac{dm_i(s)}{ds}.Z$

and all these forms satisfy the required form for a DTM subject to the conditions stated.

Theorem 7.4 If $M(s)$ is a DTM and is written in the form $P_M(s)/p_M(s)$, where $P_M(s)$ is a polynomial matrix and $p_M(s)$ is a scalar polynomial, then $P_M(s)$ is a DTM.

Proof: $P_M(s) = M(s).p_M(s) = Y.\text{diag}\,\{p_M(s)\,m_i(s)\}.Z$ and is real since M and p_M are real. The proof is clearly reversible.

Theorem 7.5 If $M(s)$ is a DTM of order p, then its elements are linear combinations of at most p linearly independent, rational functions of s.

Proof: On evaluating the triple product of eqn. 7.9, every element of $M(s)$ is some linear combination, with numeric coefficients, of the p $m_i(s)$ and every $m_i(s)$ is a rational function of s by the definition of a DTM. Some of the numeric coefficients may be zero or the $m_i(s)$ may not be l.i., so that some or all of the elements *may* be expressible as linear combinations of a smaller number of the $m_i(s)$.

Note that the converse is *not* necessarily true. For instance the matrix $\begin{bmatrix} s+1 & s+2 \\ s+3 & s+4 \end{bmatrix}$ is *not* a DTM although its elements are clearly linear combinations of the rational functions s^0 and s^1.

7.2.2.2

Theorems 7.2 to 7.5 give the essential properties of DTM's. No theorem, however, gives a defining property: they may be used, therefore, to show that a given matrix is not a DTM but not to show that it is. The one positive test remains: can the matrix be expressed in the triple product form of eqn. 7.9?

7.2.3 Application to control system design

If the plant transfer matrix $P(s)$ is a DTM, which requires $m = p$ in Fig. 5.1, the first stage in design is to imagine the plant preceded by a controller of transfer matrix $P^{-1}(c)$, where c satisfies Theorem 7.2. This combination, which we consider as a modified plant, will have a transfer matrix $Q(s) = P(s)P^{-1}(c)$ which, by Theorem 7.2, has rational eigenvalues. After this initial stage, the design is completed along the same lines as for the commutative controller, as follows.

If $P = Y.\text{diag}\,p_i.Z$, then $Q = Y\,\text{diag}\,\{p_i(s)/p_i(c)\}.Y^{-1} \equiv Y.\text{diag}\,q_i.Y^{-1}$. We then constrain $K(s)$, preceding Q, to be of the form $Y.\text{diag}\,k_i(s).Y^{-1}$ and constrain the feedback matrix $H(s)$ to be of the form $Y.\text{diag}\,h_i(s).Y^{-1}$. Thus with Y replacing $E(s)$ and $Q(s)$ replacing $P(s)$, we find, using eqns. 7.3 and 7.4, that $F(s) = Y.\text{diag}\,f_i(s).Y^{-1}$ where

$$f_i(s) = \frac{q_i k_i}{1 + q_i k_i h_i}$$

Although, after the first stage of design, the method is identical with the commutative controller, the dyadic method has two major advantages. First it deals with a plant matrix which necessarily has rational eigenvalues, thus avoiding criticism (ii) of Section 7.1; secondly, the replacement of the s-dependent $\boldsymbol{E}(s)$ by the numeric $\boldsymbol{Y}$ may be expected, in general, to give forms of $\boldsymbol{K}(s)$ and $\boldsymbol{H}(s)$ which are easier to realise physically. Note that the overall controller transfer matrix in the dyadic analysis is $\boldsymbol{P}^{-1}(c).\boldsymbol{K}(s)$ and not $\boldsymbol{K}(s)$, since $\boldsymbol{K}(s)$ is the controller for the modified plant $\boldsymbol{Q}(s)$.

Such advantages as the method possesses are, however, entirely dependent upon the plant matrix being a DTM. If this is not the case, can we approximate to the plant matrix by means of a DTM?

7.3 DTM approximations

In discussion of this problem, Owens first quotes a form suggested by Chen, which, with a slight change of notation, may be written

$$\boldsymbol{P}_C(s) \triangleq \left[\boldsymbol{P}^{-1}(0) + s\left\{\frac{d\boldsymbol{P}^{-1}(s)}{ds}\right\}_{s=0}\right]^{-1}$$

Note that, on inverting both sides, the right hand side becomes the first two terms of the Taylor expansion of $\boldsymbol{P}^{-1}(s)$ about the origin. [If $\boldsymbol{P}^{-1}(0)$ does not exist, it may be possible to use $s = c$ (real) as the basis of the expansion.] The approximation to $\boldsymbol{P}(s)$ is therefore presumably valid in the neighbourhood of $s = 0$ but may depart widely from $\boldsymbol{P}(s)$ at high values of $|s|$.

To show that $\boldsymbol{P}_C(s)$ is *in general* a DTM, write for brevity $\left\{\frac{d\boldsymbol{P}^{-1}(s)}{ds}\right\}_{s=0} \equiv \boldsymbol{T}$. Then $\boldsymbol{P}_C(s) = [\boldsymbol{P}^{-1}(0) + s\boldsymbol{T}]^{-1} = [\boldsymbol{I} + s\boldsymbol{P}(0)\boldsymbol{T}]^{-1}\boldsymbol{P}(0)$. If we suppose that the numeric matrix $\boldsymbol{P}(0)\boldsymbol{T}$ is diagonalisable and has an eigenvector assembly matrix $\boldsymbol{E}$, then $\boldsymbol{P}(0)\boldsymbol{T} = \boldsymbol{E}.\text{diag}\,\mu_i.\boldsymbol{E}^{-1}$, where the μ_i are the eigenvalues of $\boldsymbol{P}(0)\boldsymbol{T}$. Hence

$$\begin{aligned}\boldsymbol{P}_C(s) &= [\boldsymbol{I} + s\boldsymbol{E}.\text{diag}\,\mu_i.\boldsymbol{E}^{-1}]^{-1}\boldsymbol{P}(0) = [\boldsymbol{E}(\boldsymbol{I} + s\,\text{diag}\,\mu_i)\boldsymbol{E}^{-1}]^{-1}\boldsymbol{P}(0)\\ &= \boldsymbol{E}.\text{diag}\frac{1}{1+s\mu_i}.\boldsymbol{E}^{-1}\boldsymbol{P}(0)\end{aligned}$$

which is clearly of DTM form since $\boldsymbol{E}$ and $\boldsymbol{E}^{-1}\boldsymbol{P}(0)$ are both non-singular.

Owens is not satisfied with the Chen approximation on the grounds that if $\boldsymbol{P}(s)$ itself is a DTM, $\boldsymbol{P}_C(s)$ is in general different from $\boldsymbol{P}(s)$, whereas it would seem reasonable to demand their equality in this case.

Indeed, suppose that $P(s)$ is a DTM and $P(s) = Y.\text{diag}\, p_i(s).Z$. Then

$$T = \left\{\frac{dP^{-1}(s)}{ds}\right\}_{s=0} = Z^{-1}.\text{diag}\left\{\frac{dp_i^{-1}(s)}{ds}\right\}_{s=0}.Y^{-1} \equiv Z^{-1}.\text{diag}\, t_i.Y^{-1} \text{ say.}$$

Then

$$P_C(s) = [I + sP(0)T]^{-1}P(0) = [I + sY.\text{diag}p_i(0)\, t_i.Y^{-1}]^{-1}.Y\text{diag}p_i(0).Z$$

$$= Y.\text{diag}\left\{\frac{p_i(0)}{1 + sp_i(0)\, t_i}\right\}.Z$$

On comparing this with $P(s)$, we note that the two are only equal if $p_i^{-1}(s) = p_i^{-1}(0) + s\left\{\frac{dp_i^{-1}(s)}{ds}\right\}_{s=0}$, i.e. if $p_i^{-1}(s)$ is equal to the first two terms of its Taylor expansion about $s = 0$, that is, if every $p_i(s)$ is the reciprocal of a linear function of s.

Owens accordingly develops his own DTM approximation, $P_0(s)$, as follows. Compute the matrix $P(0)P^{-1}(s) \equiv A(s)$ and let $B \equiv \left\{\frac{dA(s)}{ds}\right\}_{s=0}$. Let B be diagonalisable and have an eigenvector assembly matrix F. Then

$$P_0(s) \triangleq \left[\sum_{j=1}^{p} D_j.P(s)P^{-1}(0).D_j\right]P(0)$$

where D_j is the dyad (jth column of F).(jth row of F^{-1}).

It may be shown that $P_0(s)$, like $P_C(s)$, has the property that as $s \to 0$, $P_0(s)$ only differs from $P(s)$ by a matrix whose elements are of the order of s^2. It may also be shown, that, unlike $P_C(s)$, if $P(s)$ itself is a DTM, then $P_0(s) = P(s)$.

7.3.1 Comments

(i) Owens rightly points out that any design of $K(s)$ and $H(s)$ based upon an *approximation* to the plant matrix leads, when K and H are used with the *true* plant matrix $P(s)$, to a system of which the performance must be carefully checked. Indeed it is quite possible to visualise a case where, since the approximations are only valid at low frequencies, the system (K, H, P_C) or (K, H, P_0) may be quite stable but the system (K, H, P) is unstable: this would be due to the neglect of the higher powers of s in the approximations, which may be of paramount importance in determining the behaviour of the system near its critical frequencies. If the check should prove unsatisfactory, presumably some other method of controller design must be used.

(ii) There are of course other approximations which might be attempted, not necessarily confined to low values of s. The difficulty is, however, to systematise the approximation process and especially to prove that the approximate form is indeed a DTM.
(iii) The proposed method, as put forward in this paper, pays no attention to the question of interaction.

7.4 References

1 MACFARLANE, A.G.J.: 'Commutative controller: A new technique for the design of multivariable control systems', *Electron Lett.*, 1970, **6**, pp. 121–123
2 LAYTON, J.M.: 'Commutative controller: a critical survey', *ibid.*, 1970, **6**, pp. 362–363
3 MACFARLANE, A.G.J.: 'Commutative controller: a critical survey' (author's reply), *ibid.*, 1970, **6**, pp. 363–364
4 OWENS, D.H.: 'Dyadic approximation method for multivariable control system analysis with a nuclear-reactor application', *Proc. IEE*, 1973, **120** (7), pp. 801–809
5 OWENS, D.H.: 'Dyadic expansion for the analysis of linear multivariable systems', *ibid.*, 1974, **121**, (7), pp. 713–715

7.5 Examples 7

1. Consider the real matrix $\boldsymbol{M}(s) = \begin{bmatrix} a_1 s + b_1 & a_2 s + b_2 \\ a_3 s + b_3 & a_4 s + b_4 \end{bmatrix}$ with det $\boldsymbol{M}(s) \not\equiv 0$. If det $\boldsymbol{M}(s) = fs^2 + gs + h$, find f, g, h. Show that, provided that det $\boldsymbol{M}(s)$ has distinct zeros, $\boldsymbol{M}(s)$ is always a DTM and may be expanded as $\boldsymbol{M}(s) = (p_1 s + q_1)\boldsymbol{D}_1 + (p_2 s + q_2)\boldsymbol{D}_2$, where $\boldsymbol{D}_1$ and $\boldsymbol{D}_2$ are numeric dyads and where $s = -q_1/p_1$, $s = -q_2/p_2$, are the distinct zeros of det $\boldsymbol{M}(s)$, either of which may be zero or infinite. Give the values of $\boldsymbol{D}_1$ and $\boldsymbol{D}_2$.

Show, however, that if the zeros of det $\boldsymbol{M}(s)$ coincide (whether zero, finite or infinite), certain conditions (other than $g^2 = 4fh$) must be satisfied by the coefficients in $\boldsymbol{M}(s)$ if $\boldsymbol{M}(s)$ is to be a DTM.

2. A plant has the transfer matrix $\boldsymbol{P}(s) = \dfrac{1}{(s+1)(s+2)} \begin{bmatrix} s+1 & s+2 \\ s-5 & 2s+2 \end{bmatrix}$

Confirm that this is a DTM and give its expansion. Using the DTM method, design a numeric preplant controller matrix, so that, when used with a $(-\boldsymbol{I}_2)$ feedback matrix, all natural modes of the closed-loop system are at least as damped as exp $(-3t)$.

Using such a controller, find the output vector $\boldsymbol{V}(s)$; deduce $\boldsymbol{v}(t)$ if (*a*) $R_1(s) = 1/s$, $R_2(s) = 0$, (*b*) $R_1(s) = 0$, $R_2(s) = 1/s$.

If, using the notation of the text, the pre-$Q(s)$ part of the controller is $K = Y.\text{diag}\, k_i.Y^{-1}$, obtain values of k_1 and k_2 not exceeding about 50 which will secure (i) reasonable accuracy in the ultimate response in both channels, (ii) negligible interaction, and (iii) the absence of excessive transients in either channel. (The total transients in the excited channel should at no time appreciably exceed unity, those in the unexcited channel should not appreciably exceed 0·5).

3. Verify that the plant matrix $P(s) = \dfrac{1}{(s+1)(s+2)}\begin{bmatrix} 2s+3 & s-3 \\ s-1 & 2s+5 \end{bmatrix}$ has eigenvalues which are rational in s. Try and design a satisfactory controller for this plant (assuming $-I_2$ feedback matrix) using the commutative controller technique. Noting that $P(s)$ is also a DTM, use, as an alternative, the DTM technique. (The interpretation of 'satisfactory' is left to the designer. An unexpected difficulty may occur in connection with the commutative controller technique.)

Chapter 8

Rosenbrock's inverse Nyquist array method

8.1 Introduction

This method of design for multivariable control systems[1] was developed primarily as an attempt to extend to such systems the well-tested graphical method of classical theory, the Nyquist locus technique. A desirable preliminary to such an extension is the disentanglement of the many interacting loops present in most multivariable systems and their replacement by a number of noninteracting or nearly noninteracting loops to each of which the Nyquist techniques may be applied separately. The inverse Nyquist array (INA) method of design leans heavily upon the theory of diagonally dominant matrices: the essential properties of this class of matrices will therefore be dealt with first.

8.2 Diagonally dominant matrices

Definition 8.1 A square matrix A of order n, in the field of complex numbers, is diagonally dominant (d.d) if for every i in $1 \leqslant i \leqslant n$, and every v in this range except $v = i$,

either $$|a_{ii}| > \sum_{v} |a_{ij}| \qquad (8.1a)$$

or $$|a_{ii}| > \sum_{v} |a_{ji}| \qquad (8.1b)$$

If eqn. 8.1a is satisfied, the modulus of every leading diagonal element exceeds the sum of the moduli of the other elements in the same row: then A is said to be diagonally dominant row-wise (d.d.r.). Similarly if eqn. 8.1b) is satisfied, A is said to be diagonally dominant column-wise (d.d.c.).

Definition 8.2 A square matrix $A(s)$ of order n, in the field of rational functions of the complex variable s, is diagonally dominant on a contour C in the plane of s if it is either d.d.r. *or* d.d.c. for every value of s on C.

Although this definition is applicable to the field of complex rational functions of s, we shall only be concerned with the field of *real* rational functions. Also we may make the fairly obvious but useful deduction that if $A(s)$ is d.d. on C, so is $f(s)A(s)$, where $f(s)$ is any real rational function of s, provided that no zeros or poles of $f(s)$ lie on C. For changing A into fA multiplies both sides of the inequalities by the normally positive quantity $|f|$; but if $|f| = 0$, we then have $0 > 0$, which is false, whereas if $|f| \to \infty$, we have $+\infty > +\infty$, which is unverifiable!

We shall now associate these definitions with an important theorem.

8.2.1 Theorem 8.1 (Gershgorin's)

If A is a square matrix of order n, in the field of complex numbers, its eigenvalues lie, in the complex plane, in the union of the n circles having having their centres at a_{ii} and having radii $\sum_j |a_{ij}|$. Moreover, since the eigenvalues are unchanged by transposing A, the eigenvalues also lie within the union of the circles having their centres at a_{ii} and having radii $\sum_j |a_{ji}|$ (i and j constrained as above).

8.2.1.1 Notes

(i) It will be convenient to refer to the first set of circles in the theorem as the row-set of Gershgorin circles and to the second set as the column-set of Gershgorin circles.

(ii) If A is a diagonal matrix, the a_{ii} are themselves the eigenvalues of A and all radii are zero. The theorem is therefore a statement of the limits of the area of the complex plane within which the eigenvalues are confined when nondiagonal elements are added to a diagonal matrix.

(iii) Note however that the theorem does *not* state that one particular eigenvalue remains within one particular Gershgorin circle as these nondiagonal elements are increased in modulus; for, as this occurs, the circles of either set increase in radius and at some stage overlap, allowing an eigenvalue to migrate from the inside of one circle to the inside of another (and possibly the *out*side of the first). To take a simple example in the field of real numbers, if $A = \begin{bmatrix} 24 & -16 \\ 5 & 32 \end{bmatrix}$, the eigenvalues are found to be $(28 \pm 8j)$ and are therefore both at the same distance, $4\sqrt{5}$, from both circle centres, a_{11} at (24, 0) and a_{22} at (32, 0): since $16 > 4\sqrt{5} > 5$, it follows that both eigenvalues lie *within* the circle of radius 16 but both lie *outside* the circle of radius 5, whether the row-set or column-set are considered.

(iv) Space considerations forbid a general proof of the theorem. Note, however, with $n = 2$, that the characteristic equation of A is $(\lambda - a_{11})(\lambda - a_{22}) = a_{12}a_{21}$; hence $|\lambda - a_{11}||\lambda - a_{22}| = |a_{12}||a_{21}|$ must be satisfied by each eigenvalue. If $|\lambda - a_{11}| < |a_{12}|$, then $|\lambda - a_{22}| > |a_{21}|$, but if $|\lambda - a_{11}| > |a_{12}|$ then $|\lambda - a_{22}| < |a_{21}|$: in either case the eigenvalue lies either within the circle of radius $|a_{12}|$ with centre at a_{11} or within the circle of radius $|a_{21}|$ with centre at a_{22}. In the critical case $|\lambda - a_{11}| = |a_{12}|$, then also $|\lambda - a_{22}| = a_{21}$ and the eigenvalues lie on the periphery of both these circles. This proves the theorem for $n = 2$.

8.2.1.2 Gershgorin bands

If $A(s)$ is in the field of real rational functions of s, and s describe some simple contour C in the s-plane, then, in general, both the centre of a Gershgorin circle, $a_{ii}(s)$, and its radius, $\sum_j |a_{ij}(s)|$ or $\sum_j |a_{ji}(s)|$, will vary with s. As s describes C once, the assembly of points in the complex plane on or within the successive positions of a particular Gershgorin circle as s goes round C is called a *Gershgorin band*, though 'band' is not always a suitable name. Indeed if C is, say, the circle $|s| = 1$ and if, with $n = 2$, the first row of A is, say, $[s + 1, k]$, so that the radius of the Gershgorin row-circle is constant at $|k|$ while its centre traces the circle with centre at (1, 0) and radius unity, then clearly if $|k| \geqslant 1$, the Gershgorin 'band' is the circle, centre at (1, 0), radius $(1 + |k|)$; if $|k| < 1$, the Gershgorin band is the ring-space between two circles centered on (1, 0) and of radii $(1 + |k|)$ and $(1 - |k|)$.

Note that since A is in the field of *real* rational functions of s, $a_{ii}(s^*) = a_{ii}^*(s)$, the asterisk denoting the conjugate complex. On the other hand the radius of a Gershgorin circle, being a sum of moduli, will be unchanged by changing s into s^*. Hence if the contour C is symmetrical about the real axis of s, every $a_{ii}(s)$ will map C into a contour which is symmetrical about the real axis of the complex plane and every Gershgorin band will also be symmetrical about this real axis. Note also that even if C is a simple (i.e. non-selfintersecting) contour, the mapping of C through any $a_{ii}(s)$, depending upon the degree of its numerator and denominator polynomials, need *not* be a simple contour.

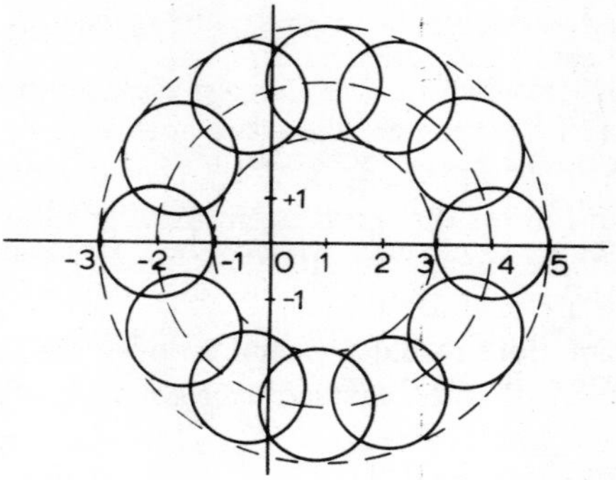

Fig. 8.1 Gershgorin band *encircling* but not *including* the origin

8.2.2 Geometrical interpretation of diagonal dominance

If we now suppose that $A(s)$ is diagonally dominant on C, the implication is that *no Gershgorin circle and therefore no Gershgorin band includes the origin of the complex plane*. since, for any value of s on C, $|a_{ii}(s)|$ is the distance from the origin of the centre of a Gershgorin circle, and, either by eqn. 8.1*a* or eqn. 8.1*b*, this distance is greater than the radius of the circle: hence the origin cannot lie within the circle and hence cannot lie within the associated band. Note, however, that this does not prevent the area of a band from *encircling* the origin: taking again a simple example, if C is the circle $|s| = 3$ and, with $n = 2$, the first row of $A(s)$ is, say $[s + 1, 1]$, then no Gershgorin circle for this row will ever include the origin (i.e. this row is d.d. on C) but the band associated with these circles clearly encircles the origin (see Fig. 8.1). In such a case, moreover, the number of encirclements of the origin by the band is necessarily equal to that made by the circle centre, or indeed by any point which, as s varies, is constrained to lie within the Gershgorin circle. Unfortunately, as noted in Note (iii) above, the eigenvalues of $A(s)$ are not so constrained and we can therefore at present make no statement about the encirclements of the origin made by the eigenvalue loci as s moves round C.

The question is made more complex by the fact that although $A(s)$ lies in the field of rational functions of s, its eigenvalues, or some of them, may be irrational functions of s. If such is the case it may indeed happen that the mapping of C through some irrational eigenvalue $\lambda_i(s)$ is not a closed contour but an open curve, so that the number of encirclements of the origin by such a curve appears difficult to assess. This particular difficulty will be surmounted by defining the number of encirclements of the origin by the locus of $\lambda_i(s)$ as the increase in $\frac{1}{2\pi}$ arg $\lambda_i(s)$ when s traces C in a prescribed direction, whether or not this increase is a whole number of revolutions. With this definition we shall show, through theorems attributable to Rosenbrock, that the total number of encirclements of the origin made by the n eigenvalues of $A(s)$ is in fact equal to the total number of encirclements of the origin made by the centres of the Gershgorin circles. We first require a topological theorem.

8.2.2.1 Theorem 8.2

Let O and P be distinct points in a plane. Let C be any contour in the plane, intersecting itself an arbitrary number of times, but not passing through O. Let Q be a point on C which describes one circuit of C in a specified direction, as a result of which arg $\overrightarrow{OQ}$ increases by $2n\pi$, (i.e. C makes n anticlockwise encirclements of O where n is some integer, positive, negative or zero). Let Q be joined to the fixed point P by a curve c in the plane, possibly also selfintersecting, the profile of which varies continuously as Q moves round C, but returns to the original profile after one circuit of C. Then whatever the position of P with respect to C, when Q describes one circuit of C, the curve c passes through O $(|n| + 2N)$ times, where N is some non-negative integer. (see Fig. 8.2)

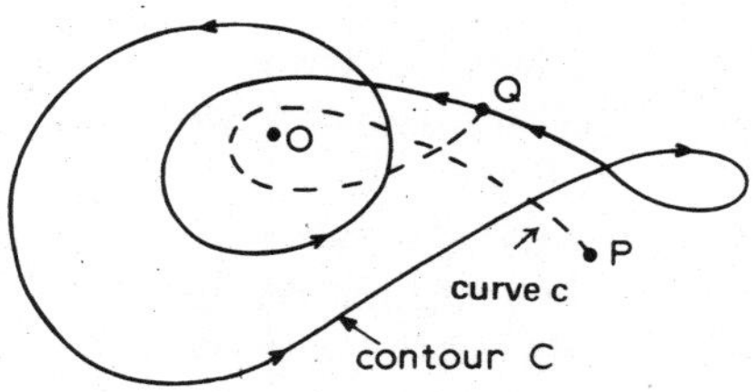

Fig. 8.2 Possible diagram to illustrate Theorem 8.2

Proof: We give a proof by physical interpretation. Consider the curve c as a string of variable length and profile, connecting P and Q and lying in the plane of C. Moreover, let us first separate the process of Q going round C from the process of c passing through O, by assuming that at O there is a peg, perpendicular to the plane of C, and therefore preventing the string c from crossing O. Take Q once round C in the prescribed direction. At any stage, the number of anticlockwise encirclements of the point O by the string PQ (considered as a locus from P to Q) may be represented by $\frac{1}{2\pi}(\arg \overrightarrow{OQ} - \arg \overrightarrow{OP})$. As Q describes C once, $\arg \overrightarrow{OQ}$ by supposition increases by $2n\pi$ and $\arg \overrightarrow{OP}$ is of course constant. Hence, to restore the string to its original profile, we shall have to lift $|n|$ turns of string over the peg, i.e. c must be made to cross O $|n|$ times; but during this process we may of course take the string to *and* fro over the peg as many times as we please, say N times. Thus c must cross O $(|n| + 2N)$ times where N is a non-negative integer.

Corollary If c does *not* pass through O as Q describes C, then $n = 0$. For then $|n| + 2N = 0$, and since both terms are non-negative, each must be zero.

We may now proceed to a more important theorem.

8.2.2.2 Theorem 8.3

Let $A(s)$ be a square matrix of order n in the field of rational functions of s. Let C be a simple contour in the s-plane such that

(*a*) $A(s)$ is diagonally dominant (row- or column-wise) on C;

(*b*) no pole of any $a_{ii}(s)$, $i = 1, 2, \ldots, n$, lies on C.

Let $\det A(s)$ map C into $C_{\det A}$ and let $a_{ii}(s)$ map C into $C_i (i = 1, 2, \ldots, n)$. Then the sum of the encirclements of the origin made by the several C_i equals the number of encirclements of the origin (in the same sense) made by $C_{\det A}$.

Proof: Note first that since both $\det A(s)$ and the several $a_{ii}(s)$ are all rational in s, both $C_{\det A}$ and the several C_i are re-entrant after one circuit of C.

Write $A \equiv A_d + A_{nd}$, where $A_d = \operatorname{diag} a_{ii}$ and A_{nd} is therefore the value of A when all its leading diagonal elements are made zero. Consider the matrix $A(k, s) = A_d + kA_{nd}$, with $0 \leqslant k \leqslant 1$, and the associated rational function of s,

$$f(k, s) \equiv \frac{\det A(k, s)}{\det A_d(s)} = \frac{\det \{A_d(s) + kA_{nd}(s)\}}{\det A_d(s)}$$

of which the denominator, $\det A_d(s)$, is equal to $\prod_{i=1}^{n} a_{ii}(s)$.

We make a number of preliminary deductions:

(i) Due to (*a*) above, no pole of any $a_{ij}(s)$, $i \neq j$, can lie on C. Therefore, using (*b*) above, all elements of A are finite on C; so therefore are all elements of $A(k, s)$; so therefore are $\det A(s)$ and $\det A(k, s)$.

(ii) Using (*a*) again, there can be no zero of any $a_{ii}(s)$ on C; hence det $A_d(s)$ cannot vanish on C; hence, using (i), $f(k, s)$ is finite on C.

(iii) Using (*a*) again, either the Gershgorin row-set or column-set does not include the origin; hence the eigenvalues of $A(s)$, which always lie within each set, cannot vanish on C. *A fortiori* the eigenvalues of $A(k, s)$, and therefore det $A(k, s)$, cannot vanish on C, since the Gershgorin circles for $A(k, s)$ have the same centres but $k(\leqslant 1)$ times the radius of those for $A(s)$. Hence $f(k, s)$ cannot vanish on C.

Now let $f(1, s) = \det A(s)/\det A_d(s)$ map C into C_f. Then by (iii) C_f cannot cross the origin 0. Also $f(0, s) = 1$ for all s. Thus for any given s on C but variable k, $f(k, s)$ will be representable by some continuous curve c joining the fixed point P or $(1, 0)$, when $k = 0$, to some point Q on C_f fixed by the particular value of s selected. As Q travels round C_f, moreover, the curve c can never cross 0, for this would imply that for some s on C and some k in the range $0 \leqslant k \leqslant 1$, $f(k, s)$ vanishes, which, by (iii), is not possible. Hence, by the corollary to the last theorem, C_f, which replaces the C of that theorem, encircles the origin zero times. Hence arg $f(1, s)$ undergoes zero change as s traces C. Hence arg det $A(s)$ and arg det $A_d(s) = \sum_{i=1}^{n} \arg a_{ii}(s)$ undergo the same change, i.e. the number of encirclements of the origin made by $C_{\det A}$ is equal to the sum of the encirclements of the origin made by the several C_i, which proves the theorem.

Since the determinant of any matrix equals the product of its eigenvalues, and therefore arg det $A(s) = \sum_{i=1}^{n} \arg \det \lambda_i(s)$, it follows that the total number of encirclements of the origin is the same for the eigenvalue loci as for the Gershgorin circle-centre loci; we shall however prefer to use the determinantal form of the theorem, as stated, rather than to introduce the eigenvalue loci, if only for two reasons: firstly because it is preferable to deal with one mapping, $C_{\det A}$, rather than n mappings, one for each $\lambda_i(s)$, and secondly because if $n > 4$, the characteristic equation for the eigenvalues of $A(s)$ is in general insoluble analytically anyway!

We should like, however, to interpret encirclements of the origin by various mappings of C in terms of the location of the zeros and poles of the relevant functions. We therefore introduce a familiar theorem.

8.2.2.3 Theorem 8.4

If $f(s)$ is a rational function of s and if C is a simple contour in the s-plane such that $f(s)$ has no poles or zeros on C, then as s traces C once in a *clockwise* direction, arg $f(s)$ increases by

$2\pi(\mathcal{P} - \mathcal{Z})$ where $\mathcal{P}, \mathcal{Z}$ are respectively the number of simple poles and simple zeros of $f(s)$ lying within C. (A pole or zero of multiplicity m counts as m simple poles or zeros.)

Proof: Suppose $f(s) = \prod_{i=1}^{m} (s - s_i)/\prod_{j=1}^{n} (s - s_j)$, where neither the s_i nor the s_j need be distinct. It will be clear from a simple vector diagram that when s traces C once clockwise, arg $(s - s_k)$ will decrease by 2π if $s = s_k$ lies within C but suffers no net change if $s = s_k$ lies outside C. Hence arg $f(s)$ will decrease by 2π for every simple zero within C and increase by 2π for every simple pole within C and is not affected by poles and zeros outside C, which proves the theorem.

The quantity $(\mathcal{P} - \mathcal{Z})$ may conveniently be called *the polar excess of $f(s)$ within C*, denoted here by $\mathcal{E}$. If then arg $f(s)$ increases by $2\pi\mathcal{E}$, this is equivalent to saying that the mapping of a clockwise tracing of C through $f(s)$ encircles its origin $\mathcal{E}$ times anticlockwise. Linking this statement with Theorem 8.3 therefore leads to the re-phrasing of this theorem as:

> If $A(s)$ is square, of order n, in the field of rational functions of s, and is diagonally dominant (d.d.r. or d.d.c.) on a simple contour C in the s-plane then, if no $a_{ii}(s)$ has any poles on C, the polar excess of det $A(s)$ within C equals the sum of the polar excesses of the $a_{ii}(s)$ within C, i.e. $\mathcal{E}_{\det A} = \sum_{i=1}^{n} \mathcal{E}_{a_{ii}}$.

We are now in a position to apply the theory developed above to the problem of controller design.

8.3 Application to controller design

8.3.1 The system considered

We visualise a system of the general lay-out of Fig. 5.1 but with the following constraints:

(*a*) $m = p$, so that $\boldsymbol{R}(s)$, $\boldsymbol{E}(s)$, $\boldsymbol{U}(s)$ and $\boldsymbol{V}(s)$ are all of order $p \times 1$ and the matrices $\boldsymbol{K}(s)$, $\boldsymbol{P}(s)$ and $\boldsymbol{H}(s)$ are all square of order p

(*b*) $\boldsymbol{K}(s)$ and $\boldsymbol{P}(s)$ are such that det $\boldsymbol{K}(s) \not\equiv 0$, det $\boldsymbol{P}(s) \not\equiv 0$, and the poles of both $\boldsymbol{K}(s)$ and $\boldsymbol{P}(s)$ lie in the open left half-plane of s

(*c*) $\boldsymbol{H}(s)$ is in general equal to $\boldsymbol{I}_p$; however, to investigate the integrity of the system when any of the feedback paths is accidentally broken, $\boldsymbol{H}$ will be considered as a possibly depleted form of the identity matrix $\boldsymbol{I}_p$, in which any number of the diagonal elements may be made zero instead of unity.

We shall make use of the following relations obtained in Chapter 5:

eqn. 5.9:
$$\frac{\text{closed-loop modal polynomial}}{\text{broken-loop modal polynomial}} = \frac{\det \boldsymbol{\Delta}(s)}{\det \boldsymbol{\Delta}(\infty)} \tag{8.2}$$

where $\boldsymbol{\Delta}(s)$ = return difference matrix

$$= \boldsymbol{I} + \boldsymbol{G}(s)\boldsymbol{H} = \boldsymbol{I} + \boldsymbol{P}(s)\boldsymbol{K}(s)\boldsymbol{H} \tag{8.3}$$

eqn. 5.2*a*;

$\boldsymbol{F}(s)$ = closed-loop transfer matrix

$= \{\boldsymbol{I} + \boldsymbol{G}(s)\boldsymbol{H}\}^{-1}\boldsymbol{G}(s) = \boldsymbol{\Delta}^{-1}(s)\boldsymbol{G}(s)$ so that

$$\boldsymbol{\Delta}(s) = \boldsymbol{G}(s)\boldsymbol{F}^{-1}(s) \text{ and hence } \det \boldsymbol{\Delta}(s) = \det \boldsymbol{G}(s)/\det \boldsymbol{F}(s) \tag{8.4}$$

eqn. 5.2*b*: $\boldsymbol{F}^{-1}(s) = \boldsymbol{G}^{-1}(s) + \boldsymbol{H}$; following a suggestion of Rosenbrock, we write the inverse of any matrix $\boldsymbol{M}$ as $\hat{\boldsymbol{M}}$ rather than $\boldsymbol{M}^{-1}$: this notation removes any ambiguity in the symbol m_{ij}^{-1}, which might, and still does, mean the reciprocal of m_{ij} but might also be interpreted as the element (i, j) of $\boldsymbol{M}^{-1}$, which will now be denoted by $\hat{m}_{ij}$. Hence the above equation will be written

$$\hat{\boldsymbol{F}}(s) = \hat{\boldsymbol{G}}(s) + \boldsymbol{H} \tag{8.5}$$

8.3.2 Stability criteria

Since $\boldsymbol{P}$ and $\boldsymbol{K}$ have their poles in the open left-half-plane (o.l.h.p.) and $\boldsymbol{H}$ is numeric, the poles of the loop-gain matrix, $-\boldsymbol{PKH}$ or $-\boldsymbol{KHP}$ or $-\boldsymbol{HPK}$, also lie in the open left-half-plane. Hence the broken-loop modal polynomial (b.l.m.p.) has its zeros in the open left-half-plane. Hence the closed-loop system is stable if and only if, using eqn. 8.2, all zeros of det $\boldsymbol{\Delta}(s)$ lie in the o.l.h.p. (One has to tread warily, for in cerain cases the closed-loop modal polynomial (c.l.m.p.) and the broken-loop modal polynomial (b.l.m.p.) have a common zero, which, if it has the same multiplicity in both, or a higher multiplicity in the b.l.m.p., will not be a zero of det $\boldsymbol{\Delta}(s)$; if this zero should have a non-negative real part, the closed-loop system might be unstable even though the zeros of det $\boldsymbol{\Delta}(s)$ are all in the o.l.h.p.. In the present context, however, this cancellation of an 'unstable' zero is impossible since the zeros of the b.l.m.p. are all 'stable', i.e. in the o.l.h.p.)

Next, owing to the simplicity of the $\hat{\boldsymbol{F}}, \hat{\boldsymbol{G}}.\boldsymbol{H}$ relation, see eqn. 8.5, as compared with the $\boldsymbol{F}, \boldsymbol{G}, \boldsymbol{H}$ relation of eqn. 5.2, we rewrite eqn. 8.4 as

$$\det \boldsymbol{\Delta}(s) = \det \hat{\boldsymbol{F}}(s)/\det \hat{\boldsymbol{G}}(s)$$

We select a contour C in the s-plane consisting of the imaginary axis from $s = -jr$ to $s = +jr$ together with a closing semicircle of radius r in the right half-plane, r being large enough to include any right half-plane poles and zeros of relevant functions. Then since det $\mathbf{\Delta}(s)$ has all its poles in the o.l.h.p. (because these are the zeros of the b.l.m.p. or a subset thereof) and since det $\mathbf{\Delta}(s)$ has all its zeros in the o.l.h.p. if the closed-loop system is stable, it follows that $\mathscr{E}_{\det \Delta} = 0$ and hence $\mathscr{E}_{\det \hat{F}} = \mathscr{E}_{\det \hat{G}}$ (within C). Moreover this is *only* true if the closed-loop system is stable. Hence a *necessary and sufficient condition of closed-loop stability is that the mappings of C through det $\hat{F}$ and det $\hat{G}$ should encircle the origin the same number of times.*

So far we have had no recourse to diagonal dominance at all. But suppose now that we have been able to choose $K(s)$ in such a way that both $\hat{G}(s)$ and $\hat{F}(s)$ are diagonally dominant (d.d.) on C. (This problem is dealt with later.) This of course means, from an operational point of view, that we are placing a certain limitation on the degree of interaction in the system, which is normally beneficial; from an analytical point of view, it means that we may make use of the final form of Theorem 8.3 given in Section 8.2.2.3. Replacing $A(s)$ in that final form by $\hat{F}(s)$ and $\hat{G}(s)$ in turn, we deduce: *if $\hat{F}(s)$ and $\hat{G}(s)$ are d.d. on C, a necessary condition for closed-loop stability is that* $\sum_{i=1}^{n} \mathscr{E}_{\hat{f}_{ii}} = \sum_{i=1}^{n} \mathscr{E}_{\hat{g}_{ii}}$ *within C.*

Moreover it follows at once from eqn. 8.5 that $\hat{f}_{ii} = \hat{g}_{ii} + h_{ii}$ and, due to the assumed structure of H, every h_{ii} is either zero or unity. If the closed-loop system is to be stable whatever feed-back links are made or broken, then considering the case $h_{ii} = 1, h_{jj}(j \neq i) = 0$, we deduce, for *any i*,

$$\mathscr{E}_{1+\hat{g}_{ii}} = \mathscr{E}_{\hat{g}_{ii}} \text{ within } C,$$

so that the number of encirclements of the origin by $\hat{g}_{ii}(s)$ and $\{1 + \hat{g}_{ii}(s)\}$ must be the same; in other words, *if $\hat{F}(s)$ and $\hat{G}(s)$ are d.d. on C, a necessary condition of closed-loop stability, whatever feedback links are made or broken, is that every $\hat{g}_{ii}(s)$ shall encircle the origin and the point* $(-1, 0)$ *an equal number of times as s traces C.* To comply with the conditions of Theorem 8.3, it is furthermore required that no $\hat{g}_{ii}(s)$ should have any poles on C, i.e. on the imaginary axis of s; if this condition is satisfied by the $\hat{g}_{ii}(s)$ it is also automatically satisfied by the $\hat{f}_{ii}(s)$, which, for any i, have the same poles.

This delightfully simple graphical criterion of system stability, coupled with complete system integrity with respect to feedback linkage faults, only requires the plotting of the p loci of the $\hat{g}_{ii}(s)$. If we choose to make r, the radius of C, tend to infinity, these loci are the Nyquist loci of the diagonal elements of $\hat{G}(s)$, the inverse of $G(s)$.

An equivalent analytical criterion, based on the fact that the poles of $\hat{g}_{ii}(s)$ and $\{1 + \hat{g}_{ii}(s)\}$ are identical, is clearly that the number of zeros of these two functions in the right-half-plane of s must be the same. This can readily be checked by using a Hurwitz criterion on the numerator polynomials of the two functions.

We are however still left with the problem of choosing $K(s)$ so that $\hat{G}(s)$ and $\hat{F}(s)$ are diagonally dominant on C.

8.3.3 The structure of the controller

Rosenbrock proves, in an Appendix to the paper, that any $\boldsymbol{K}(s)$, constrained as assumed above by the fact that det $\boldsymbol{K}(s) \not\equiv 0$ and the poles of $\boldsymbol{K}(s)$ lie in the o.l.h.p., constrained still further by forcing the zeros of det $\boldsymbol{K}(s)$ also to lie in the o.l.h.p., can always be expressed as the triple product $\boldsymbol{K}(s) \equiv \boldsymbol{K}_a . \boldsymbol{K}_b(s) . \boldsymbol{K}_c(s)$, the properties of the three factors being as follows:

$\boldsymbol{K}_a$ is a so-called commutation matrix, being null except for one element in every row and column, which is unity. Its determinant is equal to $\pm$ 1. Its inverse is its transpose, also a commutation matrix. When *pre*multiplying another matrix, it alters the order of the *rows* of that matrix; when *post*multiplying another matrix, it alters the order of the *columns*. Its purpose, in the controller, is to ensure that $R_i(s)$ affects, dominantly, $V_i(s)$ and not some $V_j(s)$.

$\boldsymbol{K}_b$ (s), when premultiplying another matrix, represents a sequence of elementary row operations, each consisting of the addition of some multiplier of one row to another. It is is desired to add $m(s)$ times row j to row i, the required matrix is the identity matrix with the addition of the element $m(s)$ in the position (i, j). The inverse of this elementary matrix is identical with it except that the sign of $m(s)$ is reversed. Since $\boldsymbol{K}_b(s)$ and $\hat{\boldsymbol{K}}_b(s)$ consist of products of such elementary matrices and since the determinant of each is clearly unity, it follows that det $\boldsymbol{K}_b(s) =$ det $\hat{\boldsymbol{K}}_b(s) = 1$. If $K_b'(s)$, the transpose of $K_b(s)$, *post*multiplies another matrix, the effect is to perform the same operations as before but this time on the *columns* of this other matrix.
The poles of every $m(s)$ multiplier lie in the open left half-plane.
$\boldsymbol{K}_c(s)$ is diagonal with no zero diagonal elements: every diagonal element has its zeros and its poles in the open left-half-plane. If we write $\boldsymbol{K}_c(s) =$ diag $k_i(s)$, it follows that $\hat{\boldsymbol{K}}_c(s) =$ diag $k_i^{-1}(s) =$ diag $\hat{k}_i(s)$ has the same properties.

It is to be noted that since $\boldsymbol{K} = \boldsymbol{K}_a\boldsymbol{K}_b\boldsymbol{K}_c$ and $\boldsymbol{G} = \boldsymbol{PK}$, $\boldsymbol{G} = \boldsymbol{PK}_a\boldsymbol{K}_b\boldsymbol{K}_c$. Hence $\hat{\boldsymbol{G}} = \hat{\boldsymbol{K}}_c\hat{\boldsymbol{K}}_b\hat{\boldsymbol{K}}_a\hat{\boldsymbol{P}}$.

8.3.3.1 Steps in the controller design

Since the criteria evolved in Section 8.3.2 are concerned with $\hat{G}(s)$ and not $G(s)$, the first step required is to find the value of $\hat{\boldsymbol{P}}(s)$, the inverse of the plant matrix. We next have to decide whether to operate with the commutation matrix $\hat{\boldsymbol{K}}_a$ or to replace it by $\boldsymbol{I}_p$, i.e. no commutation of rows. This must be decided by inspection: if in any row a particular element appears to be larger in modulus on C than the others, it should, if possible, be brought into the leading diagonal position by commutating rows; if there is no such evidence of dominance, it is probably better, as a first attempt at any rate, to put $\hat{\boldsymbol{K}}_a = \boldsymbol{I}_p$ and proceed to the next stage, premultiplication by $\hat{\boldsymbol{K}}_b(s)$.

This choice of $\hat{\boldsymbol{K}}_b(s)$ is carried out, as indicated above, by carrying out on $\hat{\boldsymbol{K}}_a\hat{\boldsymbol{P}}(s)$ a number of elementary row operations, the object of these being to make the resulting matrix d.d., usually diagonally dominant row-wise (d.d.r.), on C. The designer must here be guided by the form of the elements of the matrix and if p is at all large, much trial and error work may be involved. The work is probably best carried out by seeking to make *one row at a time* diagonally dominant, each such row being left unchanged in subsequent processes. Unless there appear to be sound reasons to do otherwise, it is worth trying a purely numerical value for $\hat{\boldsymbol{K}}_b$. The establishment, at any trial stage, of the diagonal dominance of a row requires the evaluation of the moduli of all its elements at a number of values of s on C: in simple elements this may be done analytically but for more complicated forms a computer is required and may, in fact, be developed to display a Nyquist plot for each element in turn, from which the moduli may be assessed. But it seems likely that if p is at all large the labour involved in finding a suitable $\hat{\boldsymbol{K}}_b(s)$ is very onerous.

It must be remembered also that, if possible, $\hat{\boldsymbol{K}}_b(s)$ should not merely make $\hat{\boldsymbol{K}}_b\hat{\boldsymbol{K}}_a\hat{\boldsymbol{P}}$ diagonally dominant row-wise, but also $(\hat{\boldsymbol{K}}_b\hat{\boldsymbol{K}}_a\hat{\boldsymbol{P}} + \boldsymbol{H})$; and finally the diagonal elements of $\hat{\boldsymbol{K}}_b\hat{\boldsymbol{K}}_a\hat{\boldsymbol{P}}$ must map C into loci which encircle the origin and the point $(-1, 0)$ an equal number of times. These additional requirements may be assisted by finally introducing the diagonal $\hat{\boldsymbol{K}}_c(s)$.

It is to be noted that if, for brevity, we denote $\hat{\boldsymbol{K}}_b\hat{\boldsymbol{K}}_a\hat{\boldsymbol{P}} \equiv \boldsymbol{M}$, then if $\boldsymbol{M}$ is d.d.r. so is $\hat{\boldsymbol{K}}_c\boldsymbol{M} = \hat{\boldsymbol{G}}$. For any element $\hat{k}_i$ of $\hat{\boldsymbol{K}}_c$ multiplies *all* the elements of the ith row of $\boldsymbol{M}$ by the same factor $\hat{k}_i$. This same multiplication may, however, have an effect upon the diagonal dominance of the rows of $(\boldsymbol{M} + \boldsymbol{H})$, this matrix now becoming $(\hat{\boldsymbol{K}}_c\boldsymbol{M} + \boldsymbol{H})$. As far as the encirclements criterion is concerned, it must be remembered that the zeros and poles of all $\hat{k}_i$ lie in the open left-half-plane. It follows that the mappings of C through m_{ii} and through $\hat{k}_i m_{ii}$ make an equal number of encirclements of the origin; but the mappings of $(1 + m_{ii})$ and $(1 + \hat{k}_i m_{ii})$ may make a different number of encirclements of the origin, since the two functions, although they have identical right-half-plane (r.h.p.) poles, may differ in their number of r.h.p. zeros.

On the other hand, if $\boldsymbol{K}_b$ has been found so that $\boldsymbol{M}$ and $(\boldsymbol{M} + \boldsymbol{H})$ are both d.d.r. and so that the diagonal elements of $\boldsymbol{M}$ meet the encirclements criterion, then the several $k_i(s)$ may be used to modify the dynamic properties of the dominant (diagonal) loops of the system to meet various performance requirements.

We give an example taken from Rosenbrock's paper with subsequent comments on both the example and the solution.

8.4 Example

$$\boldsymbol{P}(s) = \frac{1}{(1+s)^2}\begin{bmatrix} 1-s & 2-s \\ \frac{1}{3}-s & 1-s \end{bmatrix}, \text{ hence } \hat{\boldsymbol{P}}(s) = 3(1+s)\begin{bmatrix} 1-s & s-2 \\ s-\frac{1}{3} & 1-s \end{bmatrix}.$$

Note that $\hat{\boldsymbol{P}}(s)$ is not d.d.r. [or diagonally dominant, column-wise (d.d.c.)] at the origin. Since all the elements of the matrix part of $\hat{\boldsymbol{P}}$ are similar, linear functions of s, there does not seem to be much point in introducing the commutation matrix $\hat{\boldsymbol{K}}_a$, which we therefore equate to $\boldsymbol{I}_2$. The author at this stage suggests a premultiplier which is the inverse of $\hat{\boldsymbol{P}}(s)$ at the origin, namely $\boldsymbol{P}(0)$, with the idea of forming a product which is at any rate d.d. in the neighbourhood of the origin. We deduce $\hat{\boldsymbol{G}}_1(s)$ = interim value of $\hat{\boldsymbol{G}}(s) = \boldsymbol{P}(0)\hat{\boldsymbol{P}}(s)$

$$= (1+s)\begin{bmatrix} 1+3s & -3s \\ 2s & 1-2s \end{bmatrix}$$

The first row of $\hat{\boldsymbol{G}}_1(s)$ is d.d. on C whatever the value of its radius r. But C must include the r.h.p. zero of $\hat{g}_{1,22}$ at $s = \frac{1}{2}$ so that $r > \frac{1}{2}$; then, at $s = r$ on C the second row is *not* d.d. To remedy this, leave the first row alone and subtract m times the first row from the second row; we obtain

$$\hat{\boldsymbol{G}}_2(s) = \begin{bmatrix} 1 & 0 \\ -m & 1 \end{bmatrix}\hat{\boldsymbol{G}}_1(s) = (1+s)\begin{bmatrix} 1+3s & -3s \\ -m-(3m-2)s & 1+(3m-2)s \end{bmatrix}$$

For the second row to be d.d. at the origin requires $|m| < 1$. If $(3m-2) < 0$, we have the same difficulty as before with an r.h.p. zero of $\hat{g}_{2,22}$: hence $m > 2/3$. We compromise at $m = 5/6$, so that

$$\hat{G}_2(s) = (1+s)\begin{bmatrix} 1+3s & -3s \\ -(5+3s)/6 & (s+2)/2 \end{bmatrix}$$

Both rows are now d.d. on C whatever its radius, as may be confirmed by a simple vector diagram in the s-plane. Neither the diagonal elements nor $\det \hat{G}_2(s) = (1+s)^3$ have any r.h.p. poles or zeros, so that from this point of view r is unrestricted. The loci of the diagonal elements $\hat{g}_{2,11} = (1+s)(1+3s)$ and $\hat{g}_{2,22} = (1+s)(1+s/2)$, corresponding to s on C, are easily sketched and both encircle the origin and the point $(-1, 0)$ zero times. This is to be expected since the four functions $g_{2,11}, g_{2,22}, 1+g_{2,11}, 1+g_{2,22}$ have neither poles nor zeros in the r.h.p.

We find, however, with $\hat{F}_2 \equiv \hat{G}_2 + H$, that the first row of $\hat{F}_2$ is only diagonally dominant if $|1+(1+s)(1+3s)| > |3s(1+s)|$ which, with $s = j\omega$ on the diameter of C, is found to be satisfied only if $|\omega| < 2/\sqrt{5}$. Similarly the second row is only d.d. if $|\omega| < \sqrt{119/5}$, so that to satisfy both these conditions requires $r < 2/\sqrt{5}$. Since, to meet the other conditions, there is no restriction on r, we may make $r < 2/\sqrt{5}$ and meet all the design requirements, but it is to be noted that it is sometimes necessary to place an upper limit on r in order to meet the design requirements.

If we now introduce, for simplicity, a numerical $\hat{K}_c = \begin{bmatrix} \hat{k}_1 & 0 \\ 0 & \hat{k}_2 \end{bmatrix}$ and write $\hat{G} = \hat{K}_c \hat{G}_2$, then, as argued above, $\hat{G}$ is d.d.r. because $\hat{G}_2$ is. Provided k_1 and k_2 are positive, the diagonal elements of $\hat{G}$ will meet the encirclements criterion. As far as $\hat{F} = \hat{G} + H$ is concerned we find that its first row is diagonally dominant on C for all r only if $\hat{k}_1 > 6$, otherwise we require $r^2 < (1+\hat{k}_1)^2/\hat{k}_1(6-\hat{k}_1) = (1+k_1)^2/(6k_1 - 1)$. The second row, similarly, is d.d. for all r only if $\hat{k}_2 > 36/11$, otherwise we require $r^2 < (\hat{k}_2+6)(11\hat{k}_2+6)/\hat{k}_2(36-11\hat{k}_2) = (6k_2+1)(6k_2+11)/(36k_2 - 11)$. (It would be a useful exercise for the reader to verify these results.) As before, since no other constraints on r exist, these limits may always be met: the point is that *a* value of r may be found such that all conditions are satisfied. It is of course desirable to make the dominant loop gain-factors k_1 and k_2 appreciably larger than the critical values 1/6 and 11/36 given above, if only to increase the steady-state accuracy of the dominant loops. Rosenbrock investigates more fully a system with $k_1 = k_2 = 50$. Taking these values,

$\hat{K}(s) = \dfrac{1}{50} I_2 \cdot \begin{bmatrix} 1 & 0 \\ -\frac{5}{6} & 1 \end{bmatrix} \cdot P(0) = \dfrac{1}{300}\begin{bmatrix} 6 & 12 \\ -3 & -4 \end{bmatrix}$ so that

$K(s) = 25\begin{bmatrix} -4 & -12 \\ 3 & 6 \end{bmatrix}$. The performance of the system, if judged unsatisfactory in any respect after simulation tests or analytical investigation, may be further modified by introducing an additional $\hat{K}_c(s)$ diagonal factor, this time s-dependent, provided that we ensure that this does not invalidate any of the three design conditions: $\hat{G}(s)$ to be d.d.r., $\hat{F}(s) = \hat{G}(s) + H$ to be d.d.r., all $\hat{g}_{ii}(s)$ to encircle the origin and $(-1, 0)$ an equal number of times.

8.4.1 Comments

(i) The example is of course greatly simplified by the fact that $p = 2$ and also by the fact that $\hat{P}(s)$ is a polynomial matrix so that its elements and determinant have no finite poles: since $\hat{K}$ is numeric the same is true of $\hat{G}_1(s)$, $\hat{G}_2(s)$ and $\hat{G}(s)$. Nevertheless the example does deal with a plant transfer matrix of which the elements have r.h.p. zeros, a situation which sometimes gives rise to difficulties in design.
(ii) The controller is designed in a perfectly systematic manner; even the purely numerical controller suggested, requiring only amplifiers for its live elements, is shown to be reasonably satisfactory from the points of view of steady-state accuracy, initial transient behaviour, small degree of interaction, complete integrity in the event of feedback failures etc.
(iii) The title of the method throws perhaps an undue emphasis on Nyquist techniques. In the example the only 'Nyquist' loci required in the solution are those for the diagonal elements of $\hat{G}_2(s)$. Rosenbrock defines the Inverse Nyquist Array as the p^2 mappings of C through the several $\hat{g}_{ij}(s)$, but only those through the $\hat{g}_{ii}(s)$ appear useful and even these may be dispensed with: for the encirclements of the origin by the mapping of C through $\hat{g}_{ii}(s)$ are equal in number to the polar excess of $\hat{g}_{ii}(s)$ within C, which may easily be obtained by a Routh table for the numerator and denominator of $\hat{g}_{ii}$. In any case a Nyquist locus is normally associated with $r \to \infty$, which, as shown in the solution, may be a dangerous assumption (which is why we wrote 'Nyquist' above.)
(iv) The author's unorthodox step of premultiplying $\hat{P}(s)$ by $P(0)$ is, whatever its merits, unnecessary. Indeed the solution obtained for $\hat{K}(s)$ is $\hat{K}(s) = \begin{bmatrix} \hat{k}_1 & 2\hat{k}_1 \\ -\hat{k}_2/2 & -2\hat{k}_2/3 \end{bmatrix}$ which may be factorised either as

$$\hat{K}(s) = \begin{bmatrix} k_1 & 0 \\ 0 & k_2/3 \end{bmatrix} \begin{bmatrix} 1 & 0 \\ -3/2 & 1 \end{bmatrix} \begin{bmatrix} 1 & 2 \\ 0 & 1 \end{bmatrix} \text{ or as}$$

$$\hat{K}(s) = \begin{bmatrix} -k_1/2 & 0 \\ 0 & -2k_2/3 \end{bmatrix} \begin{bmatrix} 1 & -4 \\ 0 & 1 \end{bmatrix} \begin{bmatrix} 1 & 0 \\ 3/4 & 1 \end{bmatrix}$$

showing that the same controller may be obtained by two simple row operations, with a final stage, as before, of introducing a diagonal $\hat{K}_c$.
(v) Even in this simple example it was shown that, if $\hat{F}(s)$ was to be d.d.r., it was necessary to limit r. Yet C is to include all r.h.p. zeros and poles not merely of the $\hat{g}_{ii}(s)$ and the $\hat{f}_{ii}(s)$ but also of det $\hat{F}(s)$ and det $\hat{G}(s)$. In the example there were none; but in other cases where r may be restricted and r.h.p. poles and zeros exist, these must be found if we are to know whether the contour C, of restricted radius, includes them. Now the zeros of det $\hat{F}(s)$ are the poles of det $F(s)$ and these in turn, apart from pole-zero cancellation, are the poles of $F(s)$ itself, which determine closed-loop stability. Having found $\hat{F}(s)$ is it not possibly quicker to invert and find the poles of $F(s)$ directly as a check of closed-loop stability? The question seems worthy of consideration.
(vi) The paper implicitly gives the impression that, subject to the stated constraints on $P(s)$ and $K(s)$, the method suggested always leads to a solution satisfying the three key conditions. We are not aware of any proof of this 'existence theorem', but we are assured that no practical system has yet failed to yield to treatment!

8.5 Examples 8

1. With reference to the worked example of Section 8.4 in which $\hat{P}(s) = (1+s)\begin{bmatrix} 3(1-s) & 3(s-2) \\ 3s-1 & 3(1-s) \end{bmatrix}$, assume that $\hat{K}_a$ is the permutation matrix $\begin{bmatrix} 0 & 1 \\ 1 & 0 \end{bmatrix}$ and deduce the value of $\hat{K}_a\hat{P}(s) \equiv \hat{G}_1(s)$. Proceed now with the same methods of designing $\hat{K}_b$ as are used in the worked example and show that the introduction of $\hat{K}_a$ as above makes no difference to the controller design.
2. The Inverse Nyquist Array (i.n.a.) method of design, as presented, suggests that it is always possible, given a real, rational matrix $\hat{P}(s)$, to find a numeric real matrix $\hat{K}$ such that $\hat{K}\hat{P}(s)$ is diagonally dominant, row- or column-wise, for any s on C. The purpose of this example is to disprove this statement.

(i) Find the necessary and sufficient conditions that a matrix of form $\begin{bmatrix} a_1s+a_2 & b_1s+b_2 \\ c_1s+c_2 & d_1s+d_2 \end{bmatrix}$ shall be diagonally dominant (*a*) row-wise, (*b*) column-wise, for every s on C (assuming the radius of $C \rightarrow \infty$).

(ii) If the matrix $\begin{bmatrix} s+1 & 2(s-2) \\ 2(s+2) & 2(3s+4) \end{bmatrix}$ is premultiplied by a real, numeric matrix $\begin{bmatrix} a & b \\ c & d \end{bmatrix}$, prove that it is *not* possible to find values of a, b, c, d, which will make the product either row-diagonally-dominant or column-diagonally-dominant for all s on C,

(iii) Identifying the given matrix in (ii) with the inverse of a plant transfer matrix, find this plant transfer matrix and deduce that the plant itself is stable.

3. (On interaction). Using

$$\hat{G}_2(s) = \frac{1+s}{6}\begin{bmatrix} 6(1+3s) & -18s \\ -(5+3s) & 3(s+2) \end{bmatrix}$$ as obtained in Section 8.4, form

$\hat{G}(s) = \hat{K}_c\hat{G}_2(s)$ where $K_c = \begin{bmatrix} k_1 & 0 \\ 0 & k_2 \end{bmatrix}$ and deduce the associated value of the closed-loop transfer matrix $F(s)$ in terms of k_1 and k_2 (rather than $\hat{k}_1$ and $\hat{k}_2$). Confirm that the closed-loop modal polynomial is a stable cubic in s for any positive values of k_1 and k_2.

If $k_1 = k_2 = 10$, find the zeros of this modal polynomial, using any accepted method for solving a cubic. If the inputs $r_1(t)$, $r_2(t)$ to the closed-loop system are both unit-step functions at $t = 0$, the system being at that time inert, find the outputs $v_1(t)$ and $v_2(t)$, one mode at a time, each as the sum of two components, the first due to $r_1(t)$ and the second due to $r_2(t)$. Comment on the degree of interaction present and show, from the first part of this question, that it will be reduced by increasing k_1 and k_2.

4. MacFarlane, in his paper on the Commutative Controller, considers

$$P(s) = \frac{1}{s(s+1)^2(s+2)}\begin{bmatrix} (s+1)(2s^2+3s-1) & -(s^2+s-1) \\ (s+1)^2(s^2+s-1) & -(s+1)(s^2-2) \end{bmatrix}$$

as plant matrix with the aim of designing, with $H = I_2$, a controller $K(s)$ such that the resulting closed-loop system, when fed with unit-step input elements, shall give better than 1% ultimate error in the response without appreciable overshoot initially and with a minimum of interaction. Use the methods of this Chapter to try to design a suitable controller.

Chapter 9

Sequential design

9.1 Introduction

The principles of sequential design have been put forward by Mayne in a number of closely related papers[1,2,3,4] of which the last is probably the best introduction to the subject. In this Chapter the essentials of the method will be presented.

The method is closely associated with Rosenbrock's I.N.A. approach, dealt with in the last Chapter, but does not demand the diagonal dominance of the system transfer matrix, although it does not exclude this possibility, and therefore probably gives the designer somewhat more latitude. Moreover, as its name implies, this method of design arrives at the final result by a sequence of interdependent steps.

As with Rosenbrock's method, the control system is based upon Fig. 5.1 with $m = p$, so that $\boldsymbol{R}(s)$, $\boldsymbol{E}(s)$, $\boldsymbol{U}(s)$ and $\boldsymbol{V}(s)$ are all p-vectors and the matrices $\boldsymbol{K}(s)$, $\boldsymbol{P}(s)$ and $\boldsymbol{H}(s)$ are all square of order p; in this case, however, $\boldsymbol{H} = \boldsymbol{I}_p$ always. It is clearly desirable to restrict the poles of the controller transfer matrix $\boldsymbol{K}(s)$ to the open left-half-plane (o.l.h.p.) of s and assume for simplicity (although this is not essential) that the poles of the plant transfer matrix $\boldsymbol{P}(s)$ also all lie in the o.l.h.p. Still following Rosenbrock, if we make the further assumption that the zeros of det $\boldsymbol{K}(s)$ also lie in the o.l.h.p., we may then, as in the last Chapter, factorise $\boldsymbol{K}(s)$ as

$$\boldsymbol{K}(s) = \boldsymbol{K}_a \,.\, \boldsymbol{K}_b(s) \,.\, \boldsymbol{K}_c(s)$$

where $\boldsymbol{K}_a$ is a commutation matrix, which alters the order of the columns of any matrix which it postmultiplies and det $\boldsymbol{K}_a = \pm 1$

$\boldsymbol{K}_b$ represents a sequence of elementary column operations performed on any matrix which $\boldsymbol{K}_b$ postmultiplies, each such

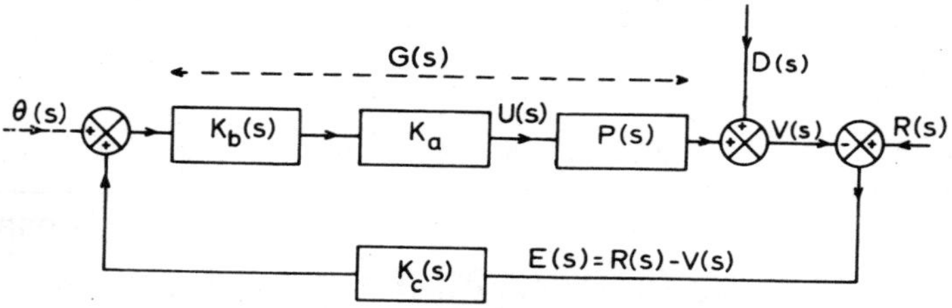

Fig. 9.1 Modified representation of control system

operation consisting of adding some multiple $m(s)$ of one column to another, where $m(s)$ has its poles in the o.l.h.p.: det $K_b(s) = 1$

K_c = diag $k_r(r = 1, 2, \ldots p)$, of which the elements $k_r(s)$ are nonzero and have all their poles and zeros in the o.l.h.p.

Since in this case $H = I_p$, it will be found more convenient to redraw Fig. 5.1 as Fig. 9.1. In this diagram the additional input vector $\boldsymbol{\theta}(s)$ is a completely mythical input, introduced merely to specify certain transfer matrices, whereas the additional input $D(s)$ represents a possible disturbance input, contributing directly to the output $V(s)$ of the system. This diagram forms the basis of the analysis which follows.

9.2 Basic relations

It will be convenient to introduce the following shorthand notation:

$$G(s) \triangleq P(s)K_aK_b(s) \tag{9.1}$$

(N.B. This is *not* the same G as in Chapter 8.)

$$\begin{aligned} L(s) &= \text{loop gain matrix with break between } P \text{ and } K_c \\ &= -G(s)K_c(s) \end{aligned} \tag{9.2}$$

$$\begin{aligned} \boldsymbol{\Delta}(s) &= \text{related return difference matrix} \\ &= I - L(s) \\ &= I + G(s)K_c(s) \end{aligned} \tag{9.3}$$

From the diagram, omitting the functional (s) for convenience,

$$E = R - V$$

$$V = GK_cE + G\boldsymbol{\theta} + D = GK_c(R - V) + G\boldsymbol{\theta} + D$$

Hence

$$(I + GK_c)V = GK_cR + G\boldsymbol{\theta} + D$$

or

$$V = \boldsymbol{\Delta}^{-1}GK_c . R + \boldsymbol{\Delta}^{-1}G . \boldsymbol{\theta} + \boldsymbol{\Delta}^{-1} . D \tag{9.4}$$

We deduce from this equation that

(i) F = closed-loop transfer matrix relating R to V

$$= \boldsymbol{\Delta}^{-1}GK_c = -(I-L)^{-1}L = -L(I-L)^{-1} \tag{9.5}$$

(ii) the closed-loop transfer matrix relating D to V is $\boldsymbol{\Delta}^{-1}$

(iii) the closed-loop transfer matrix relating $\boldsymbol{\theta}$ to V is $\boldsymbol{\Delta}^{-1}G$

9.3 Design objectives and associated requirements

We shall consider the following aspects of system performance:

(*a*) stability
(*b*) insensitivity to disturbance (D)
(*c*) insensitivity to parameter variation
(*d*) interaction
(*e*) accuracy

The topic of integrity will be deferred until the end of the Chapter.

(a) Stability Using eqn. 5.9, (c.l.m.p.)/(b.l.m.p.) = $\det \boldsymbol{\Delta}(s)/\det \boldsymbol{\Delta}(\infty)$, it is clear that the closed-loop modal polynomial will have its zeros in the o.l.h.p. (i.e. the systemswill be stable) if and only if det $\Delta(s)$ has all its zeros in the o.l.h.p. Even if the c.l.m.p. and b.l.m.p. have some common zeros which therefore cancel on the left hand side of eqn. 5.9, these can only be in the o.l.h.p. since we have assumed that the poles of P and K all lie in this region and the broken-loop system PK is therefore stable.

(b) Insensitivity to disturbance Clearly, from (ii) in the previous Section, this is ensured by making all the elements of $\boldsymbol{\Delta}^{-1}(s) \ll 1$, at any rate over some range of frequency, say $0 \leqslant \omega \leqslant \Omega$, (with $s = j\omega$), covering the frequency spectrum of the disturbance. To constrain each element of $\boldsymbol{\Delta}^{-1}(s)$ in this way is, however, too tedious a task and in most cases it is sufficient to apply the constraint to the norm of $\boldsymbol{\Delta}^{-1}(s)$, so that we require

$$\|\boldsymbol{\Delta}^{-1}(j\omega)\| \ll 1, 0 \leqslant |\omega| \leqslant \Omega. \tag{9.6}$$

Since this is an inequality, it is not a precise requirement: all we can say is that we should aim at making the left hand side as small as possible over the specified frequency range. In the same rather loose sense, we should aim at making $\|\boldsymbol{\Delta}(j\omega)\| \gg 1$, which, since $\boldsymbol{\Delta} = \boldsymbol{I} - \boldsymbol{L}$, implies

$$\boldsymbol{\Delta}(j\omega) = \boldsymbol{I} - \boldsymbol{L}(j\omega) \doteqdot -\boldsymbol{L}(j\omega) = \boldsymbol{G}(j\omega).\boldsymbol{K}_c(j\omega), 0 \leqslant |\omega| \leqslant \Omega \quad (9.7)$$

If these inequalities and approximate equalities are validated, the feed-backs are said to be 'tight': we return to this point in the next section.

(c) Insensibility to parameter variation Suppose that, due to *small* changes in the parameters of the loop, (whether in the plant or in the controller), $\boldsymbol{L}(s)$ is modified to $\boldsymbol{L}(s) + \delta\boldsymbol{L}(s)$ and that, as a result, $\boldsymbol{F}(s)$ is modified to $\boldsymbol{F}(s) + \delta\boldsymbol{F}(s)$. Since, from eqn. 9.5, $\boldsymbol{L} = -\boldsymbol{F}(\boldsymbol{I} - \boldsymbol{L})$, we deduce

$$\delta\boldsymbol{L} = -\delta\boldsymbol{F}.(\boldsymbol{I} - \boldsymbol{L}) + \boldsymbol{F}.\delta\boldsymbol{L} = -\delta\boldsymbol{F}.\boldsymbol{\Delta} - \boldsymbol{L}\boldsymbol{\Delta}^{-1}.\delta\boldsymbol{L}$$

Hence

$$\begin{aligned}\delta\boldsymbol{F} &= -(\boldsymbol{I} + \boldsymbol{L}\boldsymbol{\Delta}^{-1}).\delta\boldsymbol{L}.\boldsymbol{\Delta}^{-1}\\ &= -(\boldsymbol{\Delta} + \boldsymbol{L})\boldsymbol{\Delta}^{-1}.\delta\boldsymbol{L}.\boldsymbol{\Delta}^{-1}\\ &= -\boldsymbol{\Delta}^{-1}.\delta\boldsymbol{L}.\boldsymbol{\Delta}^{-1} \qquad (9.8)\end{aligned}$$

Hence in order to make $\boldsymbol{F}$ insensitive to variations in $\boldsymbol{L}$ we require once again that the elements of $\boldsymbol{\Delta}^{-1}(s)$ should be made as small as possible, though this time the requirement is not limited to a specified range of real frequencies but should be satisfied over as great a region of the s-plane as possible.

In this connection note that

$$\boldsymbol{\Delta}^{-1} = \text{adj}(\boldsymbol{I} + \boldsymbol{G}\boldsymbol{K}_c)/\det(\boldsymbol{I} + \boldsymbol{G}\boldsymbol{K}_c)$$

Due to the diagonal nature of $\boldsymbol{K}_c$, the ith column of $\boldsymbol{G}\boldsymbol{K}_c$ is k_i times the ith column of $\boldsymbol{G}$; it follows that the elements of adj$(\boldsymbol{I} + \boldsymbol{G}\boldsymbol{K}_c)$, which are the cofactors of the various elements of $(\boldsymbol{I} + \boldsymbol{G}\boldsymbol{K}_c)$, will in general contain a term involving the product of $(p-1)$ of the k_i (as well as terms containing the product of fewer of these same k_i), whereas det$(\boldsymbol{I} + \boldsymbol{G}\boldsymbol{K}_c)$ will contain a term involving the product of all p of the k_i (as well as terms involving the product of fewer k_i). Hence if we increase all the k_i *numerically*, i.e. if we associate each principal feed-back path with as large a gain factor as possible,,we shall decrease the value of every element of $\boldsymbol{\Delta}^{-1}$. This process will therefore tend to satisfy requirements (*b*) *and* (*c*) and we may identify 'tight' feed-backs with high gain-factors in the several $k_i(s)$. Note, however, that every element in $\boldsymbol{\Delta}^{-1}$ in general has poles corresponding to the zeros of det$(\boldsymbol{I} + \boldsymbol{G}\boldsymbol{K}_c)$, the closed-loop modal polynomial, and that no amount of necessarily finite scaling-up of the k_i can alter this situation (though it alters the *location* of these poles). Since for stability these poles lie in the o.l.h.p. of s, provided that they are not close to the imaginary axis there should be no major difficulty in satisfying eqn. 9.6; as far as requirement (*c*) is concerned, we note from eqn. 9.8 that, due to the iteration of the factor $\mathbf{\Delta}^{-1}$, $\delta \boldsymbol{F}$, and therefore $\boldsymbol{F} + \delta \boldsymbol{F}$, will contain the same poles as F but doubled in order; elsewhere in the s-plane increasing the gains of the k_i will diminish the effect of $\delta \boldsymbol{L}$ on $\delta \boldsymbol{F}$.

(d) Minimisation of interaction As noted in the last Chapter, the complete elimination of interaction implies the diagonalisation of $\boldsymbol{G}(s)$, i.e. $\boldsymbol{P}(s)\boldsymbol{K}_a\boldsymbol{K}_b(s) = \boldsymbol{M}(s)$, some diagonal matrix; thus $\boldsymbol{K}_a\boldsymbol{K}_b(s) = \boldsymbol{P}^{-1}(s).\boldsymbol{M}(s)$. In general, unless $\boldsymbol{P}(s)$ is very simple in form, this formula leads to an unnecessarily complex form for $\boldsymbol{K}_b(s)$ and in any case leads to an unstable $\boldsymbol{K}_b(s)$ if det $\boldsymbol{P}(s)$ has any zeros outside the o.l.h.p. of s!

If the feed-backs are tight, thus validating eqn. 9.7, then over $0 \leqslant |\omega| \leqslant \Omega$

$$\boldsymbol{F}(j\omega) = \mathbf{\Delta}^{-1}(j\omega).\boldsymbol{G}(j\omega)\boldsymbol{K}_c(j\omega) \doteqdot \boldsymbol{K}_c^{-1}(j\omega)\boldsymbol{G}^{-1}(j\omega)\boldsymbol{G}(j\omega)\boldsymbol{K}_c(j\omega)$$
$$= \boldsymbol{I}_p \qquad (9.9)$$

so that interaction should be reasonably small. On the other hand, as $s \to \infty$, $\boldsymbol{P}(s)$ normally tends to zero, and with it $\boldsymbol{G}(s)$, since $\boldsymbol{K}_a\boldsymbol{K}_b(s)$ will tend to some finite limit, zero or not. Thus to avoid interaction at high frequencies requires that as $s \to \infty$, $\boldsymbol{G}(s)\boldsymbol{K}_c(s)$, as it tends to null matrix, should approximate to a diagonal form: for since $GK_c \to 0$, $\Delta \to I_p$ so that $F \to GK_c$.

(e) Accuracy If the feedbacks are tight and eqn. 9.9 is validated, accuracy (equality of output and input) is very high; as a particular case, since $s = 0$ is in the stated range, the ultimate error (as $t \to \infty$) will be very small.

9.3.1 Summary

Summarising these results, requirements (*b*), (*c*), (*d*) and (*e*) will tend to be satisfied if we are able to design the controller with high gains associated with all the elements of $\boldsymbol{K}_c(s)$, but of course this must be achieved while meeting the essential stability requirement (*a*), that $\mathbf{\Delta}(s)$ should have all its zeros in the o.l.h.p. As in one-input/one-output systems it may be found that high gains lead to instability.

9.4 Sequential design of $K_c(s)$

Let us assume that K_a and $K_b(s)$ have been designed (we shall return to this design later) so that $G(s)$ is known. Starting with $K_c(s) = 0$, we propose to try and design its elements $k_1(s), \ldots, k_p(s)$ one at a time until, when all are present, we shall have completed the design. It will be necessary to introduce some additional symbols.

We define

$$K_{cq}(s) \triangleq \text{diag}\,\{k_1(s), k_2(s), \ldots k_q(s), 0, 0, \ldots 0\} \tag{9.10}$$

= value of K_c with only the first q elements present, the remainder being zero ($q = 0, 1, 2, \ldots, p$)

$$\boldsymbol{\Delta}_q(s) \triangleq I + G(s)K_{cq}(s) \tag{9.11}$$

= associated value of return difference matrix

$$G^q(s) \triangleq \{\boldsymbol{\Delta}_q(s)\}^{-1}G(s) \tag{9.12}$$

= associated value of the transfer matrix relating to the output $V(s)$ to the mythical input $\boldsymbol{\theta}(s)$ in Fig. 9.1.

Note that, in the symbol $G^q(s)$, q is a superfix and not a power: this is done, as will be seen, to allow more room for suffixes. Note also the particular cases: $K_{c0} = 0$; $K_{cp} = K_c$; $\boldsymbol{\Delta}_0 = I$; $\boldsymbol{\Delta}_p = \boldsymbol{\Delta}$; $G^0 = G$; $G^p = \boldsymbol{\Delta}^{-1}G$.

We shall also introduce the *scalar* return difference function

$$\delta_q(s) \triangleq 1 + k_q(s)\,g_{qq}^{q-1}(s) \tag{9.13}$$

where g_{qq}^{q-1} is, using normal notation, the qth diagonal element of G^{q-1}. The importance of this last function arises from the fact that, when we add the element $k_q(s)$ to $K_{c(q-1)}(s)$, we are introducing a feedback $\{-k_q(s)\}$ from the qth element of $V(s)$ to the qth element of $\boldsymbol{\theta}(s)$, these two quantities being, at that stage, related by the 'forward feed' $g_{qq}^{q-1}(s)$.

Finally, it will be convenient to introduce an unambiguous notation for rows and columns of a matrix; for any matrix M we define

$$m_{q.} \triangleq q\text{th row of } M, \quad m_{.q} \triangleq q\text{th column of } M,$$

a convenient notation for which we are indebted to Mayne.

9.4.1 Recurrence relations

It will be noted from eqn. 9.11 that $\mathbf{\Delta}_q$ differs from $\mathbf{\Delta}_{q-1}$ only by the term involving k_q in the product $\boldsymbol{GK}_{cq}$, i.e.

$$\begin{aligned}
\mathbf{\Delta}_q &= \mathbf{\Delta}_{q-1} + k_q \,.\, \boldsymbol{G} \,.\,(\text{null matrix except for } q\text{th diagonal element} = 1)\\
&= \mathbf{\Delta}_{q-1} + k_q \,.\, \boldsymbol{G} \,.\, \boldsymbol{i}_{.q}\boldsymbol{i}_{q.}\\
&= \mathbf{\Delta}_{q-1}\{\boldsymbol{I} + k_q \,.\, \mathbf{\Delta}_{q-1}^{-1}\, \boldsymbol{G} \,.\, \boldsymbol{i}_{.q}\boldsymbol{i}_{q.}\\
&= \mathbf{\Delta}_{q-1}\{\boldsymbol{I} + k_q \,.\, \boldsymbol{G}^{q-1} \,.\, \boldsymbol{i}_{.q}\boldsymbol{i}_{q.}\} \text{ (using eqn. 9.12)}\\
&= \mathbf{\Delta}_{q-1}\{\boldsymbol{I} + k_q \,.\, \boldsymbol{g}_{.q}^{q-1}\boldsymbol{i}_{q.}\} \qquad (9.14)
\end{aligned}$$

a relation from which we shall deduce two important results.

In the first place

$$\begin{aligned}
\det \mathbf{\Delta}_q &= \det \mathbf{\Delta}_{q-1} \,.\, \det\{\boldsymbol{I} + k_q\boldsymbol{g}_{.q}^{q-1}\boldsymbol{i}_{q.}\}\\
&= \det \mathbf{\Delta}_{q-1} \,.\, (1 + k_q g_{qq}^{q-1}) = \delta_q \det \mathbf{\Delta}_{q-1}
\end{aligned}$$

(Note that $\{\boldsymbol{I} + k_q\boldsymbol{g}_{.q}^{q-1}\boldsymbol{i}_{q.}\}$ consists of unity matrix plus only a qth column and that, as a result, the cofactor of every element in this qth column is zero except for the qth element for which the cofactor is unity.) We deduce

$$\begin{aligned}
\det \mathbf{\Delta}_q &= \delta_q\delta_{q-1} \det \mathbf{\Delta}_{q-2} = \ldots = \delta_q\delta_{q-1}\ldots\delta_1 \det \mathbf{\Delta}_0\\
&= \delta_q\delta_{q-1}\ldots\delta_1 \text{ since } \mathbf{\Delta}_0 = \boldsymbol{I} \qquad (9.15)
\end{aligned}$$

As a particular case, with $q = p$,

$$\det \mathbf{\Delta} = \det \mathbf{\Delta}_p = \delta_p\delta_{p-1}\ldots\delta_1 \qquad (9.16)$$

Since it is known that, for stability, the zeros of $\det \mathbf{\Delta}$ must all lie in the o.l.h.p. it follows that *for stability the zeros, if any, of every $\delta_q (q = 1, 2, \ldots p)$ must all lie in the o.l.h.p.*

Secondly, taking inverses of eqn. 9.14:

$$\mathbf{\Delta}_q^{-1} = \{\boldsymbol{I} + k_q\boldsymbol{g}_{.q}^{q-1}\boldsymbol{i}_{q.}\}^{-1}\,\mathbf{\Delta}_{q-1}^{-1} \qquad (9.17)$$

Noting again that the bracketed matrix consists of unity matrix plus a qth column, it is easy to show that its inverse has the same structure and *its* added qth column is easily found. For if $(\boldsymbol{I} + \boldsymbol{B})$ is the inverse of say $(\boldsymbol{I} + \boldsymbol{a}_{.q}\boldsymbol{i}_{q.})$, then $(\boldsymbol{I} + \boldsymbol{B})(\boldsymbol{I} + \boldsymbol{a}_{.q}\boldsymbol{i}_{q.}) = \boldsymbol{I}$, so that

$$\boldsymbol{B} + \boldsymbol{a}_{.q}\boldsymbol{i}_{q.} + \boldsymbol{B}\boldsymbol{a}_{.q}\boldsymbol{i}_{q.} = \boldsymbol{0}$$

If we equate the rth column of the left hand side to zero with $r \neq q$, then, noting that the rth column of $\boldsymbol{a}_{.q}\boldsymbol{i}_{q.}$ is null and that therefore so is the rth column of $\boldsymbol{Ba}_{.q}\boldsymbol{i}_{q.}$, it follows that the rth column of $\boldsymbol{B}$ is also null, so that $\boldsymbol{B}$ may be written as $\boldsymbol{b}_{.q}\boldsymbol{i}_{q.}$. We then have

$$\boldsymbol{b}_{.q}\boldsymbol{i}_{q.} + \boldsymbol{a}_{.q}\boldsymbol{i}_{q.} + \boldsymbol{b}_{.q}\boldsymbol{i}_{q.}\boldsymbol{a}_{.q}\boldsymbol{i}_{q.} = \boldsymbol{0}$$

or

$$\boldsymbol{b}_{.q}\boldsymbol{i}_{q.} + \boldsymbol{a}_{.q}\boldsymbol{i}_{q.} + \boldsymbol{b}_{.q}a_{qq}\boldsymbol{i}_{q.} = \boldsymbol{0}$$

and on equating the qth column to zero we obtain

$$\boldsymbol{b}_{.q} + \boldsymbol{a}_{.q} + a_{qq}\boldsymbol{b}_{.q} = \boldsymbol{0} \text{ so that } \boldsymbol{b}_{.q} = -\boldsymbol{a}_{.q}/(1 + a_{qq})$$

Applying this result to eqn. 9.17 after postmultiplying both sides by $\boldsymbol{G}$ leads to:

$$\boldsymbol{G}^q = \{\boldsymbol{I} - k_q g^{q-1}_{.q}\boldsymbol{i}_{q.}/(1 + k_q g^{q-1}_{qq})\}\boldsymbol{G}^{q-1}$$

$$= \{\boldsymbol{I} - k_q g^{q-1}_{.q}\boldsymbol{i}_{q.}/\delta_q\}\boldsymbol{G}^{q-1}$$

or

$$\boldsymbol{G}^q = \boldsymbol{G}^{q-1} - g^{q-1}_{.q}g^{q-1}_{q.}k_q/\delta_q \tag{9.18}$$

The sequential design of $\boldsymbol{K}_c$ may now be carried out by using eqns. 9.13 and 9.18 alternately, with gradually increasing values of q. Starting with $q = 1$ in eqn. 9.13, remembering that $\boldsymbol{G}^0 = \boldsymbol{G}$ (and is assumed known), is the basis for designing $k_1(s)$, which should be as simple as possible in form (possibly numeric) but *must* be such that the zeros of $\delta_1(s)$ lie in the o.l.h.p.; moreover, if possible, this constraint should be satisfied even if the scale of $k_1(s)$ is increased by multiplying it by some large number m_1. If such a $k_1(s)$ can be found, it determines δ_1 using eqn. 9.13, and the value of both may be substituted in eqn. 9.18, with $q = 1$, to determine $\boldsymbol{G}^1(s)$. We then repeat the cycle putting $q = 2$ in eqn. 9.13 to determine $k_2(s)$ and $\delta_2(s)$ which, in eqn. 9.18 with $q = 2$, gives $\boldsymbol{G}^2(s)$ etc. etc., until, using eqn. 9.13 with $q = p$, k_p is designed and δ_p deduced, thus completing the design of $\boldsymbol{K}_c(s)$. Use of eqn. 9.18 with $q = p$ finally gives $\boldsymbol{G}^p$, which, as noted previously, equals $\boldsymbol{\Delta}^{-1}\boldsymbol{G}$. Thus $\boldsymbol{F} = \boldsymbol{\Delta}^{-1}\boldsymbol{GK}_c$ may be calculated.

9.4.2 Some comments

Suppose that, in the above design process, for some value of $q < p$, we have got as far as finding $\boldsymbol{G}^{q-1}$, so that the next step is to use eqn. 9.13, in the form $\delta_q = 1 + k_q g^{q-1}_{qq} m_q$ (where m_q is a scaling factor) to design k_q in such a way that the zeros of δ_q lie in the o.l.h.p. over as wide a range of m_q as possible. For convenience we shall write this equation in the form

$$\delta_q(s) = \{f_1(s) + m_q f_2(s)\}/f_1(s) \equiv f(s)/f_1(s)$$

where $f_1(s)$ is a polynomial of degree n_1, comprising the polar factors of k_q and g_{qq}^{q-1};

$f_2(s)$ is a polynomial of degree n_2, comprising the zeroing factors of k_q and g_{qq}^{q-1}, the former, by supposition, lying in the o.l.h.p.

$f(s) = f_1(s) + m_q f_2(s)$, its zeros being identical with those of $\delta_q(s)$.

Note that $n_2 \leqslant n_1$ since neither k_q nor g_{qq}^{q-1} will, in practice, have more zeros than poles. Note also that the poles of k_q are, by supposition, in the o.l.h.p. and that the poles of g_{qq}^{q-1} are some selection of the poles of $\mathbf{G}^{q-1}$, that is, of the zeros of det $\boldsymbol{\Delta}_{q-1} = \prod_{r=1}^{q-1} \delta_r$, which, in the earlier stages of the design, have also been placed in the o.l.h.p.: $f_1(s)$ is therefore a Hurwitz polynomial. Our aim is to make $f(s)$ a Hurwitz polynomial over as wide a range of m_q as possible, possibly even as $m \to \infty$.

If $m_q = 0$, the zeros of $f(s)$ are the zeros of $f_1(s)$ which lie in the o.l.h.p. Using elementary root-locus theory, if $m_q \to \infty$, n_2 of the n_1 zeros of $f(s)$ tend to the zeros of $f_2(s)$ while the remaining $(n_1 - n_2)$ zeros of $f(s)$ tend to infinity along asymptotes making an angle (odd integer) $\pi/(n_1 - n_2)$ with the positive real axis of s, these asymptotes being moreover concurrent at the point on the real axis.

$$s = s_c = \frac{1}{n_1 - n_2}\left\{\sum \text{zeros of } f_1 - \sum \text{zeros of } f_2\right\}$$

We deduce the following general observations:

(*a*) if $f(s)$ is to be Hurwitz over an infinite range of m_q then $f_2(s)$ must be Hurwitz, i.e. the zeros of $g_{qq}^{q-1}(s)$ must lie in the o.l.h.p.; it is important to note that this is only a *necessary* condition and not a *sufficient* one

(*b*) if $n_1 - n_2 \geqslant 3$, it is *impossible* for $f(s)$ to be Hurwitz over the infinite range of m_q, since two or more of the asymptotes terminate in the r.h.p.; it seems probable that the range of m_q over which $f(s)$ is Hurwitz will be increased by making s_c as negative as possible, that is, by making the sum of the zeros of f_2 appreciably greater than the sum of the zeros of f_1, which in turn implies introducing phase-advance factors into k_q or g_{qq}^{q-1} or both;

(*c*) if $n_1 - n_2 = 2$, there are two collinear asymptotes parallel to the imaginary axis: it is essential in this case that s_c should be negative, but although the root-loci then start and terminate in the o.l.h.p. they may still cross over into the r.h.p. for intermediate values of m_q: e.g. $f(s) = (s^2 + s + 1)(s^2 + s + 2) + m_q(s^2 + s + 4)$ is *not* Hurwitz if $1 \leqslant m_q \leqslant 7$.

(*d*) if $n_1 - n_2 = 1$ the only asymptote is the negative real axis, but even in this case roots may migrate into the right hand side, e.g. $f(s) = (s^2 + s + 1)(s^2 + s + 2) + m_q(s^3 + s^2 + 12s + 5)$ is not Hurwitz in the approximate range $0{\cdot}37 \leqslant m_q \leqslant 12{\cdot}84$. Even when $n_1 = n_2$ the danger persists: if $f(s) = (s^3 + 6s^2 + 6s + 33) + m_q(s^3 + 3s^2 + 3s + 8)$, it is not Hurwitz in $5 + \sqrt{13} \geqslant 2m_q \geqslant 5 - \sqrt{13}$.

(*e*) If $n_1 = 2$ and f_2 is Hurwitz then f is Hurwitz over the full range of k.

In view of (*a*), it will certainly assist the design if $f_2(s)$ is a Hurwitz polynomial which means that every $g_{qq}^{q-1}(s)$ should have its zeros in the o.l.h.p. for $q = 1, 2, \ldots p$; in particular g_{11}^0 or g_{11}. To this extent at any rate the satisfactory design of $k_1(s)$ depends upon the value of $\boldsymbol{G}(s) = \boldsymbol{P}(s)\boldsymbol{K}_a\boldsymbol{K}_b(s)$, which is of course to be adjusted by a suitable choice of $\boldsymbol{K}_a$ and $\boldsymbol{K}_b(s)$. It will be noted that a further requirement on $k_1(s)$ is that when its value is substituted in eqn. 9.18 to find $\boldsymbol{G}^1(s)$, this matrix should be such that $g_{22}^1(s)$ has *its* zeros in the o.l.h.p. in order to facilitate the design of $k_2(s)$, and so on through the various stages of the design. There appears to be considerable scope for trial and error.

In view of (*b*) and (*c*) above, it would also appear that a guiding principle should be to make s_c as negative as possible in these cases.

Apart from these general guide-lines, however, it is obviously necessary, in view of the examples given above, that every candidate $k_q(s)$ function should be rigorously tested by submitting the resulting $\delta_q(s)$ to a Nyquist locus test or its numerator, $\{f_1(s) + m_q f_2(s)\}$ to a Routh-table test, in order to verify that, for sufficiently large and possibly infinite m_q, the zeros of every $\delta_q(s)$ in turn lie in the o.l.h.p. If this is the case, then the zeros of every det $\Delta_q(s)$, using eqn. 9.15, also lie in the o.l.h.p. and therefore every $\boldsymbol{G}^q(s)$, using eqn. 9.12, is a stable transfer matrix.

9.5 Design of $G(s)$

The only *explicit* constraint on $\boldsymbol{G}(s) = \boldsymbol{P}(s)\boldsymbol{K}_a\boldsymbol{K}_b(s)$ encountered in the sequential design of $\boldsymbol{K}_c$ is that $g_{11}^0(s) = g_{11}(s)$ should have its zeros in the o.l.h.p. Mayne's subsequent analysis, which we largely adhere to, is based on the more general requirement, mentioned in (*a*) above, that the zeros of every $g_{qq}^{q-1}(s)$, $q = 1, 2, \ldots p$, should lie in the o.l.h.p.; to the extent that $\boldsymbol{G}$ affects every $\boldsymbol{G}^{q-1}$, this is an *implicit* demand on $\boldsymbol{G}$.

Equating elements (*a, b*) on each side of eqn. 9.18, with $m_q k_q$ replacing k_q, gives

$$g^q_{ab} = g^{q-1}_{ab} - g^{q-1}_{aq} g^{q-1}_{qb} m_q k_q / \delta_q \quad \text{where} \quad \delta_q = 1 + m_q k_q g^{q-1}_{qq}$$
$$= \{g^{q-1}_{ab} + m_q k_q (g^{q-1}_{ab} g^{q-1}_{qq} - g^{q-1}_{aq} g^{q-1}_{qb})\}/\delta_q$$

It at once follows that if $a = q$ or $b = q$ or $a = b = q$, this simplifies to

$$g^{q-1}_{qb} = g^{q-1}_{qb}/\delta_q; \; g^q_{aq} = g^{q-1}_{aq}/\delta_q; g^q_{qq} = g^{q-1}_{qq}/\delta_q \qquad (9.19)$$

whereas if the feedback k_q is tight over some region of the s-plane so that $m_q |k_q g^{q-1}_{qq}| \gg 1$, then

$$g^q_{ab} \doteqdot (g^{q-1}_{ab} g^{q-1}_{qq} - g^{q-1}_{aq} g^{q-1}_{qb})/g^{q-1}_{qq} \; (a, b \neq q) \qquad (9.20)$$

Thus, in the sequential process, the introduction of k_q leads to a $\boldsymbol{G}^q$ of which every element in both the qth row and column is obtained by dividing the corresponding element in $\boldsymbol{G}^{q-1}$ by δ_q while leaving other elements of normal magnitude; hence, after introducing $k_1, k_2, \ldots k_p$ to form $\boldsymbol{G}^p$, any element $g^p_{ab} (a \neq b)$ will have an order of magnitude $1/\delta_a \delta_b$ whereas any element g^p_{aa} will have an order of magnitude $1/\delta_a$, so that $\boldsymbol{G}^p$ will *approximate* to a diagonal matrix (and so therefore will $\boldsymbol{F} = \boldsymbol{G}^p \boldsymbol{K}_c$). This is not unexpected since, by eqn. 9.9, with tight feedbacks, $\boldsymbol{F} \to \boldsymbol{I}$.

We can, however, go further. Suppose that in the sequential design process $\boldsymbol{G}^{q-1}$ has been obtained and k_q is to be introduced to form $\boldsymbol{G}^q$. Since $k_1, k_2, \ldots k_{q-1}$ have already been introduced, it follows that the first, second, ..., $(q-1)$th rows and columns of $\boldsymbol{G}^{q-1}$ have already been reduced by an order of magnitude $\delta_1^{-1}, \delta_2^{-1}, \ldots, \delta_{q-1}^{-1}$ but that the remaining elements of G^{q-1} are of normal magnitude. Consider in G^q the diagonal element g^q_{aa}, with $a < q$, given by eqn. 9.20 with $b = a$:

$$g^q_{aa} \doteqdot (g^{q-1}_{aa} g^{q-1}_{qq} - g^{q-1}_{aq} g^{q-1}_{qa})/g^{q-1}_{qq} \quad (a < q)$$

In this expression $g^{q-1}_{aa}, g^{q-1}_{aq}, g^{q-1}_{qa}$ are all of order of magnitude $1/\delta_a$ while g^{q-1}_{qq} is of normal magnitude. Thus if all feedbacks are tight so that $|\delta_a| \gg 1$, it follows that

$$g^q_{aa} \doteqdot g^{q-1}_{aa} \quad (a < q)$$

In other words the introduction of k_q $(q > a)$ leaves the element (a, a) unchanged, or approximately so. Noting that $g^a_{aa} = g^{a-1}_{aa}/\delta_a$ by eqn. 9.19,

$$g^p_{aa} \doteqdot g^{p-1}_{aa} \doteqdot \ldots \doteqdot g^{a+1}_{aa} \doteqdot g^a_{aa} = g^{a-1}_{aa}/\delta_a, \quad a = 1, 2, \ldots, p$$

Since, moreover, $\boldsymbol{G}^p$ approximates to a diagonal matrix, its determinant approximates to the product of its leading diagonal elements; hence

$$\det G^p \doteqdot \prod_{a=1}^{p} \{g_{aa}^{a-1}/\delta_a\}$$

or, using eqns. 9.15 and 9.12,

$$\det G = \det \mathbf{\Delta}_p \,.\, \det G^p \doteqdot \prod_{a=1}^{p} g_{aa}^{a-1}$$

Noting finally that $G = P.K_a.K_b$ and that $\det K_a = \pm 1$ and $\det K_b = 1$, we have

$$\det P(s) \doteqdot \pm \prod_{a=1}^{p} g_{aa}^{a-1}(s) \tag{9.21}$$

This equation, though approximative, is interesting, for since it is based upon all feedbacks being tight and since it was shown [deduction (*a*) in Section 9.4.2] that a *necessary* condition for tight feedbacks is that every g_{aa}^{a-1} should have its zeros in the o.l.h.p., it follows that tight feedbacks are impossible unless all zeros of $\det P(s)$ are also in the o.l.h.p. If this is *not* the case then one or more of the feedbacks will *not* be tight. Even if this *is* the case, the selection of k_1, say, determines a G^1 which may be such that the zeros of g_{22}^1 are not in the o.l.h.p., and so on throughout the sequential process. The purpose of K_a and, to a greater extent, $K_b(s)$, is, in Mayne's opinion, to 'distribute' the behaviour of the given $\det P(s)$ among the various $g_{aa}^{a-1}(s)$ so that a tight feedback design can be achieved. This concept in turn leads to the idea of a sequential design of G, synchronised, stage by stage, with the sequential design of K_c.

9.5.1 Sequential design of $G(s)$

In the first place the permutation matrix K_a should be selected so that $P(s)K_a$ is such that its diagonal elements are, *if possible*, greater in modulus than other elements in the same row or column over some region of s surrounding the origin. This may be impossible: if, with $p = 2$, $|p_{21}| > |p_{11}| > |p_{22}| > |p_{12}|$, no column-interchange nor, indeed, re-ordering of the output elements, will yield a matrix with the required characteristics: we do the best we can and possibly have alternative options to be investigated if necessary. We then have $G = (PK_a).K_b$, where (PK_a) is known and K_b is to be found sequentially. This is accomplished by considering K_b as the product of p factors:

$$K_b(s) = K_{b1}(s).K_{b2}(s) \ldots K_{bp}(s)$$

in which, as will be seen later, K_{bq} is to be selected immediately before selecting k_q, the qth element of the diagonal K_c. We suppose moreover that every K_{bq} has the structure, in partitioned form,

$$K_{bq} = \begin{bmatrix} I_{q-1} & 0 \\ 0 & Z^q \end{bmatrix}$$

where Z^q is therefore square of order $(p-q+1)$. Moreover, K_{bq} represents a sequence of elementary column operations (adding some multiple of one column to another column) performed on whatever matrix K_{bq} postmultiplies. It follows that every $\det K_{bq} = 1$. Note that, due to its structure, K_{bq} can only perform column operations on the last $(p-q+1)$ columns of the matrix it postmultiplies and that $\det Z^q = 1, K_{b1} = Z^1, K_{bp} = I_p$.

Due again to the structure of K_{bq}, it follows that

$$K_{b1} \cdot K_{b2} \ldots K_{bp} \cdot K_{cq} = K_{b1} \cdot K_{b2} \ldots K_{bq} \cdot K_{cq}$$

for it is easily shown that $K_{br}K_{cq} = K_{cq}$ provided $r \geqslant q+1$. [Remember that $K_{cq} = \text{diag}\,(k_1, k_2, \ldots k_q, 0, \ldots 0)$].

It follows that the return difference matrix $\mathbf{\Delta}_q$ defined by eqn. 9.11, may be written as

$$\mathbf{\Delta}_q = I + PK_aK_{b1}K_{b2} \ldots K_{bp} \cdot K_{cq}$$

or as

$$\mathbf{\Delta}_q = I + PK_aK_{b1}K_{b2} \ldots K_{bq} \cdot K_{cq}$$

the loop-gain being unchanged by the addition of $K_{b(q+1)}, \ldots K_{bp}$.

Thus, proceeding as in the development of eqn. 9.14,

$$\begin{aligned} \mathbf{\Delta}_q - \mathbf{\Delta}_{q-1} &= PK_aK_{b1} \ldots K_{bp}(K_{cq} - K_{c,q-1}) \\ &= k_q PK_aK_{b1} \ldots K_{bp} i_{.q} i_{q.} \end{aligned}$$

but, noting that $K_{br} i_{.q} i_{q.} = i_{.q} i_{q.}$ if $r \geqslant q+1$, this may be written

$$\begin{aligned} \mathbf{\Delta}_q &= \mathbf{\Delta}_{q-1} + k_q PK_aK_{b1} \ldots K_{bq} i_{.q} i_{q.} \\ &= \mathbf{\Delta}_{q-1}(I + \mathbf{\Delta}_{q-1}^{-1} k_q PK_aK_{b1} \ldots K_{bq} i_{.q} i_{q.}) \end{aligned}$$

At this stage we shall find it convenient to introduce (in parallel with G^q of the last section), *two* similar functions:

$$\left.\begin{aligned} \bar{G}^q &\triangleq \mathbf{\Delta}_q^{-1} PK_aK_{b1} \ldots K_{bq} \\ \text{and} \quad \tilde{G}^q &\triangleq \mathbf{\Delta}_q^{-1} PK_aK_{b1} \ldots K_{b,q+1} = \bar{G}^q . K_{b,q+1} \end{aligned}\right\} \tag{9.22}$$

We then have

$$\mathbf{\Delta}_q = \mathbf{\Delta}_{q-1}(I + k_q \tilde{G}^{q-1} i_{.q} i_{q.}) \tag{9.23}$$

or

$$\det \mathbf{\Delta}_q = \det \mathbf{\Delta}_{q-1} \cdot \det\,(I + k_q \tilde{G}^{q-1} i_{.q} i_{q.})$$

$$= \det \boldsymbol{\Delta}_{q-1} . (1 + k_q \tilde{g}_{qq}^{q-1})$$

$$= \det \boldsymbol{\Delta}_{q-1} . \tilde{\delta}_q$$

where

$$\tilde{\delta}_q = 1 + k_q \tilde{g}_{qq}^{q-1} \tag{9.24}$$

Noting that $\boldsymbol{\Delta}_0 = \boldsymbol{I}$, we deduce by iteration

$$\det \boldsymbol{\Delta}_q = \prod_{r=1}^{q} \tilde{\delta}_r$$

Hence, provided that all $\tilde{\delta}_r(s)$ have their zeros in the o.l.h.p., the same will be true of the zeros of all $\det \boldsymbol{\Delta}_q(s)$, including, as an important member, $\det \boldsymbol{\Delta}_p(s) = \det \boldsymbol{\Delta}(s)$. In other words both the completed system and the systems for which $\boldsymbol{K}_{cq}$ is substituted for $\boldsymbol{K}_c$ will be stable. This condition should be obeyed even if the various $k_q(s)$ are associated with a high, if not infinite, gain factor m_q, a *necessary* condition for this being that every $\tilde{g}_{qq}^{q-1}(s)$ should also have its zeros in the o.l.h.p.

Moreover, as in the last Section, from eqn. 9.22,

$$\boldsymbol{\Delta}_{q-1} \bar{\boldsymbol{G}}^{q-1} \boldsymbol{K}_{bq} = \boldsymbol{P}\boldsymbol{K}_a \boldsymbol{K}_{b1} \ldots \boldsymbol{K}_{bq} = \boldsymbol{\Delta}_q \bar{\boldsymbol{G}}^q$$

or, using eqn. 9.23,

$$= \boldsymbol{\Delta}_{q-1}(\boldsymbol{I} + k_q \tilde{\boldsymbol{G}}_{q-1} \boldsymbol{i}_{.q} \boldsymbol{i}_{q.}) \bar{\boldsymbol{G}}^q$$

so that, cancelling the nonsingular $\boldsymbol{\Delta}_{q-1}$ we deduce

$$\bar{\boldsymbol{G}}^q = (\boldsymbol{I} + k_q \tilde{\boldsymbol{G}}_{q-1} \boldsymbol{i}_{.q} \boldsymbol{i}_{q.})^{-1} . \bar{\boldsymbol{G}}^{q-1} \boldsymbol{K}_{bq}$$

$$= (\boldsymbol{I} - k_q \tilde{\boldsymbol{G}}^{q-1} \boldsymbol{i}_{.q} \boldsymbol{i}_{q.} / \tilde{\delta}_q) . \tilde{\boldsymbol{G}}^{q-1}$$

$$= \tilde{\boldsymbol{G}}^{q-1} - k_q \tilde{g}_{.q}^{q-1} \tilde{g}_{q.}^{q-1} / \tilde{\delta}_q \tag{9.25}$$

giving $\bar{\boldsymbol{G}}^q$ in terms of $\tilde{\boldsymbol{G}}^{q-1}$.

We may now proceed to the design. We define stage q in the design as that stage at which $k_1, k_2, \ldots k_{q-1}$ and $\boldsymbol{K}_{b1}, \boldsymbol{K}_{b2}, \ldots \boldsymbol{K}_{b,q-1}$ have been found so that $\bar{\boldsymbol{G}}^{q-1}$ is known. (Note that $\bar{\boldsymbol{G}}^0 = \boldsymbol{P}\boldsymbol{K}_a$.) To reach stage $(q + 1)$ we proceed by three steps:

(i) choose $\boldsymbol{K}_{bq}$ so that $\tilde{\boldsymbol{G}}^{q-1} = \bar{\boldsymbol{G}}^{q-1} \boldsymbol{K}_{bq}$ is such that $\tilde{g}_{qq}^{q-1}$ has its zeros in the o.l.h.p.: (this requirement involves only the first column of $\boldsymbol{Z}^q$ in $\boldsymbol{K}_{bq}$; it may be possible at the same time, by a suitable choice of the other columns, to make $\tilde{\boldsymbol{G}}^{q-1}$ approximate to an upper or lower-triangular matrix which, as Mayne shows, is conducive to integrity);

(ii) choose k_q so that $\tilde{\delta}_q$, given by eqn. 9.24, has its zeros in the o.l.h.p. for as high an associated gain-factor m_q as possible: it may be that the feasible range of m_q can be increased by using a different value of $\boldsymbol{K}_{bq}$ in (i), thus possibly requiring some trial-and-error in these two associated steps;

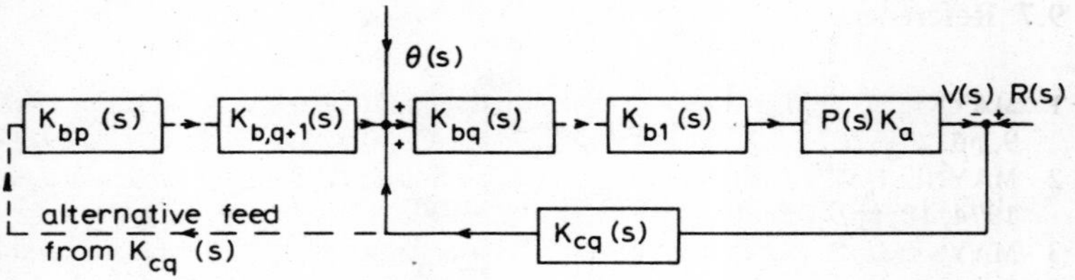

a

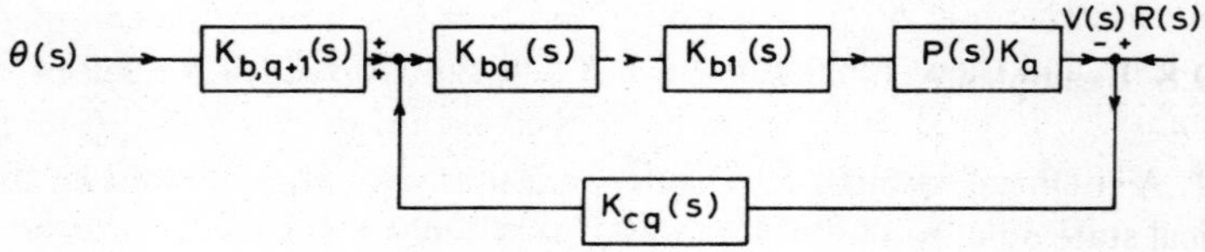

b

Fig. 9.2 Diagrammatic representation of (*a*) $\bar{G}^q(s)$, (*b*) $\tilde{G}^q(s)$
(*a*) Contribution to $V(s)$ from $\theta(s)$ is $G^q(s).\theta(s)$. This is *not* altered by using the alternative feed from $K_{cq}(s)$
(*b*) Contribution to $V(s)$ from $\theta(s)$ is $\tilde{G}^q(s)$. Here there is no alternative path

(iii) use eqn. 9.25 to calculate $\bar{G}^q$. We are now at stage $(q + 1)$ and the three steps may be repeated.

The possibly academic functions $\bar{G}^q$ and $\tilde{G}^q$ are given a more concrete significance as transfer matrices in Fig. 9.2.

9.6 A note on integrity

Provided that all the $\delta_q(s)$ in Section 9.4 or all the $\tilde{\delta}_q(s)$ in Section (9.5) have their zeros in the o.l.h.p., the complete system will be stable. It will still be stable if certain types of failure occur in the paths through K_c, notably if there is a simultaneous break in $k_{q+1}, k_{q+2}, \ldots, k_p$ for some q in $0 \leqslant q \leqslant p-1$. There is however no guarantee of stability if any other combinations of failure occur in the feedback paths through K_c, for instance if $k_1 = 0$. This topic, as well as integrity in the presence of faults in the plant, is dealt with in the papers mentioned.

9.7 References

1 MAYNE, D.Q.: 'The design of linear multivariable systems', *Automatica*, 1973, 9, pp. 201–207
2 MAYNE, D.Q.: 'The effect of feedback on linear multivariable systems', *ibid.*, 1974, **10**, pp. 405–412
3 MAYNE, D.Q., and CHUANG, S.C.: 'The sequential return difference method for designing linear multivariable systems', Proceedings of the IEE conference on computer-aided control systems, Cambridge, England, 1973
4 MAYNE, D.Q.: 'The sequential return difference design method', Proceedings of the Vacation School for SRC Postgraduate Control Engineers, UMIST Module, Manchester, England, April 1976

9.8 Examples 9

1. A chemical reactor, after linearisation, is representable in the canonical state equation form $\dot{\boldsymbol{x}} = \boldsymbol{A}\boldsymbol{x} + \boldsymbol{B}\boldsymbol{u}$, $\boldsymbol{v} = \boldsymbol{C}\boldsymbol{x} + \boldsymbol{D}\boldsymbol{u}$, by

$$\boldsymbol{A} = \begin{bmatrix} 1{\cdot}383 & 0 & 1{\cdot}343 & 0{\cdot}208 \\ -0{\cdot}446 & -4{\cdot}155 & 0 & 0{\cdot}135 \\ 5{\cdot}096 & 0 & -8{\cdot}0 & 0 \\ -1{\cdot}663 & 24{\cdot}76 & 0 & -1{\cdot}104 \end{bmatrix},$$

$$\boldsymbol{B} = \begin{bmatrix} 0 & 0 \\ 5{\cdot}679 & 0 \\ 0 & -1{\cdot}573 \\ 0 & 0 \end{bmatrix}, \quad \boldsymbol{C} = \begin{bmatrix} 1 & 0 & 0 & 0 \\ 0 & 1 & 0 & 0 \end{bmatrix}, \quad \boldsymbol{D} = \boldsymbol{0}$$

Calculate the value of the output-to-input transfer matrix, $\boldsymbol{G}(s) = \boldsymbol{V}(s)/\boldsymbol{U}(s)$, expressing $\boldsymbol{G}(s)$ as the quotient of a polynomial matrix and a scalar polynomial; deduce that the reactor is unstable on open loop.

Use the methods of sequential control design to obtain a closed-loop system, paying attention to stability, steady-state error, overshoot, under conditions of step-function input in either channel, and cross-talk. Although a purely numeric controller may be attempted in the first place, it is probable that a more satisfactory design will be obtained by using controller elements containing an integration, i.e. of the form $(a + b/s)$. (Adapted from Mayne[3])

2. Try, using the methods of the present Chapter, to design a satisfactory controller for a plant represented by

$$G(s) = V(s)/U(s)$$

$$= \frac{1}{s(s+1)^2(s+2)} \begin{bmatrix} (2s^2+3s-1)(s+1) & -2(s^2+s-1) \\ (s^2+s-1)(s+1)^2 & -(s^2-2)(s+1) \end{bmatrix}$$

(This is the matrix used by A.G.J. MacFarlane in connection with Commutative Controller Design).

Part 3

Optimisation theory

Chapter 10

Introduction to optimisation

10.1 Optimal and nonoptimal

In Part II we have analysed a number of techniques for designing multivariable controllers for specified plants such that the resulting closed-loop systems had *satisfactory* properties in certain areas of performance, e.g. stability, steady-state accuracy, absence of interaction, integrity etc. In the majority of industrial control systems, the cost of a control system (mostly capital expenditure) must be weighed against the expected benefits of its installation, such as possible reduction in production costs, higher uniformity of product, reduced likeliness of plant breakdowns etc. These economic factors therefore lead in practice to a compromise solution for the controller, which must be reasonably cheap to instal yet satisfactory, within certain margins of tolerance, in its performance.

Clearly, if the economic strait-jacket is relaxed while at the same time the tolerances in technical performance are made more stringent, there will be a trend towards designing a controller which is not merely 'satisfactory' but which is the best that can be designed whatever criterion of 'bestness' the designer has in mind. It is not surprising therefore to find that such fields of development as space travel, guided weaponry etc., (where the cost of development and production is borne by the tax-payer, while the performance of the end-product must be above reproach) should have stimulated the development of optimisation theory and that such development should have antedated the development of the 'satisfactory' techniques reviewed in Part II.

10.2 Introductory concepts

The optimisation of a quantity is understood to mean its minimisation

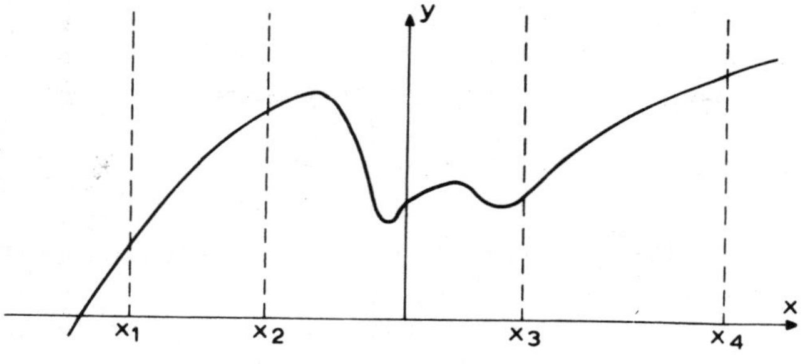

Fig. 10.1 Illustrating greatest and least values of y within some restricted range of x

(or maximisation) by the variation of variables upon which the given quantity depends. Since, algebraically, the maximisation of a quantity implies the minimisation of its negative, it will usually be convenient to identify optimisation with *min*imisation, particularly as most optimisation problems are initially concerned with minimisation (of cost, of time, of fuel consumption etc.)

The minimisation (or maximisation) of a function is often associated with equating its partial derivatives to zero. However, we must realise that even if we have set aside the economic strait-jacket, we are still constrained by physical limitations. To take a very simple instance, if we wish to take a train from rest at station A to rest at station B in minimum time, we can do this *theoretically* by imposing a very high acceleration during the first part of the journey followed by a very high deceleration during the second part; by making the acceleration and deceleration tend to inifinity we can make the time of travel (by Newtonian mechanics) tend to zero. And if Einstein forbids us to travel faster than light, other constraints will occur well before this speed is reached, for instance the maximum safe speed of the track; also an unduly high acceleration implies a very large prime mover (which increases the mass to be accelerated) while an unduly high deceleration implies a very high rate of energy dissipation (quite apart from the fact that both are in any case limited by other considerations such as the comfort of the passengers, the grip of the driving wheels, the stresses in various parts of the driving mechanism etc.) The minimisation of time from A to B must therefore be achieved within certain ranges of speed, acceleration and deceleration.

More generally consider (see Fig. 10.1) a function y given in terms of x by the graph shown. If x is constrained to the range x_1 to x_4, the minimum value of y occurs at $x = x_1$, which is not a turning value at all; its maximum occurs at $x = x_4$, which again is not a turning value.

On the other hand, if x is constrained to the range x_2 to x_3, the least value of y clearly does occur at a turning value, but there are two such 'minimum' turning values in the range; the greatest value of y is also at a turning value but there are two such 'maximum' turning values in the range. Hence the optimal point of operation, when the independent variables are limited in range, may be an *extremum* point of the range, or may be a turning value within the range and, if the latter, we may have to choose between several such points by computing the function at each point. (We have of course simplified the problem here by considering y to be a function of only one independent variable).

10.3 Mathematical classification

In the literature certain types of optimisation problems are given titles. Note that, in any case, almost all optimisation problems in the field of control consist of minimising some criterion function of the system variables by adjusting the input quantity to the system, which, as in Part I, we denote by $\boldsymbol{u}(t)$.

(i) The Lagrange problem. This may be stated as follows: given the system representation $\dot{\boldsymbol{x}} = \boldsymbol{f}(\boldsymbol{x}, \boldsymbol{u}, t)$ and $\boldsymbol{x}(t_0) = \boldsymbol{x}_0$, find $\boldsymbol{u}(t)$ if the criterion function $I = \int_{t_0}^{t_1} F(\boldsymbol{x}, \boldsymbol{u}, t)\, dt$ is to be minimised, F being a differentiable function of its arguments and t_1 given.

(ii) The Mayer problem. Given $\dot{\boldsymbol{x}} = \boldsymbol{f}(\boldsymbol{x}, \boldsymbol{u}, t)$ and $\boldsymbol{x}(t_0) = \boldsymbol{x}_0$, and given the value of *some* of the elements of $\boldsymbol{x}(t_1)$, t_1 being unspecified, find $\boldsymbol{u}(t)$ if the criterion function $J = [G(\boldsymbol{x}, \boldsymbol{u}, t)]_{t_0}^{t_1}$ is to be minimised.

(iii) The Bolza problem, which is essentially a combination of the other two: with the same data as in (ii), find $\boldsymbol{u}(t)$ if the criterion function $K = [G(\boldsymbol{x}, \boldsymbol{u}, t)]_{t_0}^{t_1} + \int_{t_0}^{t_1} F(\boldsymbol{x}, \boldsymbol{u}, t)\, dt$ is to be minimised.

It will be noted that (i) is an integral minimisation problem, that (ii) is an end-point minimisation problem and that (iii) is a combination of both. This distinction is however rather slender. Indeed we can convert an end-point problem to an integral problem by introducing the unit impulse function $\delta(t)$: using the integral properties of this function we may write

$$[G(\boldsymbol{x}, \boldsymbol{u}, t)]_{t_0}^{t_1} = \int_{t_0-}^{t_1+} G(\boldsymbol{x}, \boldsymbol{u}, t)\{\delta(t - t_1) - \delta(t - t_0)\}\, dt$$

Any objections to this transformation on the grounds that the integrand is no longer a differentiable function of t may be set aside either by replacing the δ-functions by t-differentiable 'approach functions' (which tend to an impulse function as some parameter tends to zero or infinity, for instance $\delta(t) = \text{Lt}\ \{|a|.\exp(-a^2t^2)\}/\sqrt{\pi}$ as $a \to \infty$) or alternatively by appealing to the calculus of generalised functions.

Conversely, an integral problem may be converted into a terminal value problem by the following artifice. If it is required to minimise $\int_{t_0}^{t_1} F(\boldsymbol{x}, \boldsymbol{u}, t)\,dt$ where $\boldsymbol{x}$ is of order $n \times 1$, set $x_{n+1}(t) \equiv \int_{t_0}^{t} F(\boldsymbol{x}, \boldsymbol{u}, \tau)\,d\tau$ and consider $x_{n+1}(t)$ as an additional element of the state vector, with dynamic equation $\dot{x}_{n+1} = F(\boldsymbol{x}, \boldsymbol{u}, t)$. Then we have to minimise $x_{n+1}(t_1)$, which is an end-point problem.

It will therefore be more useful to classify optimisation problems along physical or engineering (rather than mathematical) lines, bearing in mind that, as noted above, the problems may be constrained by *inequality* constraints (such as $\boldsymbol{u}_1 \leqslant \boldsymbol{u}(t) \leqslant \boldsymbol{u}_2$, by which we mean $u_{1r} \leqslant u_r(t) \leqslant u_{2r}$, $r = 1, 2, \ldots, m$; or $\boldsymbol{x}_1 \leqslant \boldsymbol{x}(t) \leqslant \boldsymbol{x}_2$, implying $x_{1r} \leqslant x_r(t) \leqslant x_{2r}$, $r = 1, 2, \ldots, n$; $t_0 \leqslant t \leqslant t_1$) as well as, of course, by the *equality constraint* of the state equation $\dot{\boldsymbol{x}} = \boldsymbol{f}(\boldsymbol{x}, \boldsymbol{u}, t)$.

10.4 Practical classification

10.4.1 Error minimisation

The error in a controlled quantity is normally defined as its required or target value minus its actual value. It is often convenient to incorporate the required values into an enlarged state vector: for instance if the required value of one of the state elements is a unit step function, add the element $x_{n+1}(t) = 1$, with $\dot{x}_{n+1} = 0$; or if a requirement is sinusoidal, say $K \sin kt$, we may introduce two additional state elements x_{n+1}, x_{n+2}, related by $\begin{bmatrix} \dot{x}_{n+1} \\ \dot{x}_{n+2} \end{bmatrix} = \begin{bmatrix} 0 & 1 \\ -k^2 & 0 \end{bmatrix} \begin{bmatrix} x_{n+1} \\ x_{n+2} \end{bmatrix}$ with initial conditions $x_{n+1}(0) = 0$ and $x_{n+2}(0) = Kk$: x_{n+1} is here the requirement, $K \sin kt$, satisfying the differential equation $\ddot{x}_{n+1} = -k^2 x_{n+1}$ while $x_{n+2} = \dot{x}_{n+1} = Kk \cos kt$. The number of additional state vector elements required depends upon the order of the differential equation satisfied by the requirement function.

The purpose of these artifices is merely to allow the error in each controlled state element to be expressed in terms of the elements of the (enlarged) state vector. The method is optional.

It will usually be desired to minimise some weighted mean value of the several errors (equal in number to the controlled quantities), weighted that is, not merely as between the several controlled quantities but also weighted with respect to time. This last concept arises from the fact that, if we are bringing the system from some arbitrary initial state to some final state, the initial errors will probably be relatively high, but not through any fault of the control system; to penalise the system for these high initial errors therefore seems illogical. But we do expect these initial errors to be attenuated as quickly as possible: it is therefore reasonable to weight the error by multiplying it by some positive, monotonically increasing function of time, before assessing its mean value in some way, over the control interval. We might therefore be led to a criterion function of Lagrangian form such as $I = \int_{t_0}^{t_1} w(t)\{k_1e_1^2 + k_2e_2^2 + \ldots\}dt$ where the k_r are positive constants, the $e_r(t)$ are the components of the error vector and $w(t)$ is the weighting function. The quadratic form in the e_r has been chosen partly because we are clearly interested in minimising the errors in *magnitude*, whether they are positive or negative, partly because the mean-square value of a time-varying quantity is a widely used form of assessing its value over a given time-interval, partly because the quadratic form is much more tractable mathematically than, say, a linear function of the moduli of the several errors.

As pointed out above, the minimisation of I may have to be achieved in the presence of constraints on the magnitudes of the input elements. Alternatively, and this is often adequate, some degree of limitation on, at any rate, the mean value of these magnitudes may be achieved by adding to the integrand of I some term such as $(K_1u_1^2 + K_2u_2^2 + \ldots +)$: the relative magnitudes of the K_r and the k_r then reflect the importance attached to minimising the input components as compared with minimising the errors.

Finally, we may attach particular importance to the values of some or all of the errors at $t = t_1$, the end-point of the control interval. If some of these are specified, then clearly we are specifying also the terminal values of some of the state-vector elements (Mayer problem). Alternatively we may, instead of specifying these terminal values, merely attach particular importance to them in the minimisation process by adding a term to I, additional to the integral, of the form $\{h_1e_1^2(t_1) + h_2e_2^2(t_1) + \ldots +\}$ (Bolza problem). [Note that this is equivalent to adding to $w(t)$, while retaining the integral form of I, terms which are suitable multiples of the impulse function $\delta(t - t_1)$.]

10.4.2 Time minimisation

The problem in this case is normally to find $\boldsymbol{u}(t)$ such that the state-vector is changed from some given initial value $\boldsymbol{x}(t_0)$ to some given final value $\boldsymbol{x}(t_1)$ in a minimum time $(t_1 - t_0)$. The problem is almost always associated with inequality constraints on the magnitude of the input-vector elements, for in the absence of such constraints the change can usually be accomplished in zero time (e.g. the train problem of Section 10.2). The problem may be treated in terms of the integral criterion function $I = \int_{t_0}^{t_1} dt$, the integrand being unity, or may be considered as the minimisation of an additional state-vector element $x_{n+1}(t) = \int_{t_0}^{t} dt$, (so that $\dot{x}_{n+1} = 1$,) at the end-point $t = t_1$.

10.4.3 Fuel minimisation

The problem is of paramount importance in connection with the analysis of rocket-propelled space-craft where, clearly, any reduction which can be made in the mass of fuel to be carried results in a possible increase in pay-load (passengers, instrumentation etc.) The rate of fuel consumption of a jet engine is some monotonically increasing function of the thrust developed, the thrust in its turn being the input to the space vehicle. Thus in changing the 'state' (i.e. position, velocity, orientation etc.) of the vehicle from one value to another, minimising fuel consumption implies minimising a criterion function of integral form, $I = \int_{t_0}^{t_1} h(u).dt$, for a single jet or a linear combination of such terms in the case of several jets, the various $h(u)$ being all monotonically increasing functions of u. (Note that if the vehicle is fitted with retro-active jets, their fuel consumption is still positive although their thrust is reversed: it is therefore more accurate to consider the fuel consumed by a jet as $\int_{t_0}^{t_1} h(|u|)dt$, irrespective of the direction of the thrust u.)

If we make the simplifying assumption that thrust is directly proportional to rate of fuel consumption, then, allowing the various jets present to have different thrust-to-consumption ratios, the criterion integral will assume the form $I = \int_{t_0}^{t_1} \{k_1|u_1(t)| + k_2|u_2(t)| + \ldots\}dt$, a Lagrange problem in which $\boldsymbol{x}(t)$ and t are absent from the integrand.

10.5 Conclusion

Other optimisation problems than these may of course occur. It must be remembered, however, that the criterion function used should always be mathematically tractable if an analytical, as opposed to a numerical or computational, solution is to be obtained. With this in mind, it is probable that the three problem types just outlined will cover the majority of cases encountered. The mathematical tools available for their solution will be described in the Chapters that follow.

Note finally this important point. All the optimisation problems described require the finding of the plant input-vector $\boldsymbol{u}(t)$ required to minimise some criterion function: we are not discussing here *how* this required input to the plant is to be generated; it may be that only in rare cases will it be possible to generate this input-vector by feedback techniques. Apart from these rare cases we are therefore concerned basically with an open-loop system (the plant) and not with closed-loop theory.

Chapter 11

Aspects of the calculus of variations

11.1 A preliminary problem

To illustrate the relevant principles of the calculus of variations, we shall consider first, in some detail, a relatively simple problem which, being 2-dimensional, may be graphically illustrated.

Problem I: In the (t, x) plane, given two nonintersecting curves $x = g_0(t), x = g_1(t), g_0$ and g_1 being differentiable, it is required to find (i) a trajectory joining (t_0, x_0) on the first curve to (t_1, x_1) on the second curve, (ii) the co-ordinates of these end-points, such that the integral along this trajectory, $x = x(t)$, given by

$$I_x = \int_{t_0}^{t_1} F(x, \dot{x}, t)\,dt$$

has a turning value, F being a differentiable function of its arguments.

Solution: Refer to Fig. 11.1. We assume in the first place that a solution exists, i.e. that there is some optimal trajectory $x = X(t)$, say, joining the points (T_0, X_0) and (T_1, X_1) such that I_X is a turning value of I_x. Consider an adjoining trajectory specified by $x(t) = X(t) + \delta X(t)$, joining $(T_0 + \Delta T_0, X_0 + \Delta X_0)$ to $(T_1 + \Delta T_1, X_1 + \Delta X_1)$, all increments being considered small quantities of the first order. Since I_X is a turning-value of I_x (note there are no inequality constraints in this problem), we may argue that the change in I_x due to changing x from X to $(X + \delta X)$ contains no terms of the first order of small quantities. But

$$\delta I_X \equiv I_{X+\delta X} - I_X = \int_{T_0+\Delta T_0}^{T_1+\Delta T_1} F(X+\delta X, \dot{X}+\delta\dot{X}, t)\,dt - \int_{T_0}^{T_1} F(X, \dot{X}, t)\,dt;$$

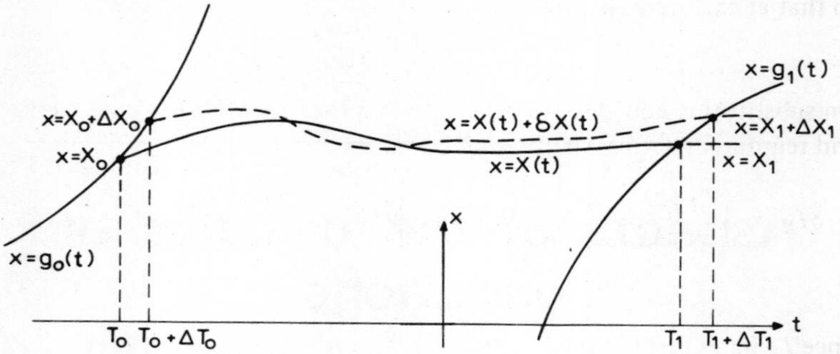

Fig. 11.1 Relating to Problem I

expanding the first integrand by a Taylor series, retaining only first-order terms gives

$$F(X+\delta X, \dot{X}+\delta\dot{X}, t) = F(X, \dot{X}, t) + \frac{\partial F(X, \dot{X}, t)}{\partial X}\delta X + \frac{\partial F(X, \dot{X}, t)}{\partial \dot{X}}\delta\dot{X}$$

Hence

$$\begin{aligned}\delta I_X &= \int_{T_0}^{T_1}\left\{\frac{\partial F}{\partial X}\delta X + \frac{\partial F}{\partial \dot{X}}\delta\dot{X}\right\}dt + \\ &\quad \left\{\int_{T_1}^{T_1+\Delta T_1} - \int_{T_0}^{T_0+\Delta T_0}\right\}\left(F + \frac{\partial F}{\partial X}\delta X + \frac{\partial F}{\partial \dot{X}}\delta\dot{X}\right)dt \\ &= \int_{T_0}^{T_1}\left\{\frac{\partial F}{\partial X}\delta X + \frac{\partial F}{\partial \dot{X}}\delta\dot{X}\right\}dt + \Delta T_1 . F_1 - \Delta T_0 . F_0\end{aligned}$$

to a first order, (11.1)

in which, for brevity, we have written F for $F(X, \dot{X}, t)$ and F_0 and F_1 for the values of this function at the optimal trajectory terminal points (T_0, X_0) and (T_1, X_1) respectively. Moreover, on integrating by parts,

$$\int_{T_0}^{T_1}\frac{\partial F}{\partial \dot{X}}\delta\dot{X}\,dt = \left[\delta X . \frac{\partial F}{\partial \dot{X}}\right]_{T_0}^{T_1} - \int_{T_0}^{T_1}\delta X . \frac{d}{dt}\left(\frac{\partial F}{\partial \dot{X}}\right)dt \qquad (11.2)$$

In addition

$$\begin{aligned}X_1 + \Delta X_1 &= \overline{X+\delta X}(T_1+\Delta T_1) = X(T_1+\Delta T_1) + \delta X(T_1+\Delta T_1) \\ &= X(T_1) + \dot{X}(T_1)\Delta T_1 + \delta X(T_1) \text{ to a first order,}\end{aligned}$$

so that

$$\Delta X_1 = \dot{X}(T_1)\Delta T_1 + \delta X(T_1);$$

but

$$\Delta X_1 = \dot{g}_1(T_1)\Delta T_1, \text{ so that } \delta X(T_1) = \{\dot{g}_1(T_1) - \dot{X}(T_1)\}\Delta T_1$$

Similarly, at the other end, $\delta X(T_0) = \{\dot{g}_0(T_0) - \dot{X}(T_0)\}\Delta T_0$

so that at each terminal point

$$\delta X = (\dot{g} - \dot{X})\Delta T \tag{11.3}$$

On substituting eqn. 11.3 in eqn. 11.2 and then eqn. 11.2 in eqn. 11.1 and regrouping terms we obtain:

$$\delta I_X = \int_{T_0}^{T_1} \left\{ \frac{\partial F}{\partial X} - \frac{d}{dt}\frac{\partial F}{\partial \dot{X}} \right\} \delta X.dt + \left[\left\{ (\dot{g} - \dot{X})\frac{\partial F}{\partial \dot{X}} + F \right\} \Delta T \right]_{T_0}^{T_1} = 0 \tag{11.4}$$

since I_X has a turning value with respect to the *variations* δX in X.

The important point to appreciate at this stage is that eqn. 11.4 must be satisfied for *every* first-order variation δX from the optimal trajectory $x = X(t)$, provided only that $\delta X(t)$ is time-differentiable.

11.1.1 The Euler-Lagrange equation

Consider, for instance, those variations which lead to the same end-points (T_0, X_0) and (T_1, X_1) as the optimal trajectory. For such variations, ΔT vanishes at both end-points so that eqn. 11.4 reduces to

$$\int_{T_0}^{T_1} \left\{ \frac{\partial F}{\partial X} - \frac{d}{dt}\frac{\partial F}{\partial \dot{X}} \right\} \delta X.dt \equiv \int_{T_0}^{T_1} \phi(t).\delta X.dt = 0 \tag{11.5}$$

Obviously we *can* satisfy this equation by making $\phi(t)$ vanish at every point of the trajectory: we propose to show that this is not merely a sufficient but also a necessary condition for satisfying eqn. 11.5. For if $\phi(t)$ is *not* zero for every t in $T_0 \leqslant t \leqslant T_1$, suppose that it is positive (or negative) in some subinterval $T_0 \leqslant a \leqslant t \leqslant b \leqslant T_1$. Now consider a variation given by

$$\begin{aligned} \delta X(t) &= k^2(t-a)^2(t-b)^2 \text{ in } a \leqslant t \leqslant b \\ &= 0 \text{ in the rest of the range.} \end{aligned}$$

Since δX is positive or zero in $T_0 \leqslant t \leqslant T_1$, the integrand of eqn. 11.5 will have the same sign as $\phi(t)$, (positive *or* negative) in $a \leqslant t \leqslant b$ and will be zero outside this subrange so that eqn. 11.5 cannot be satisfied. Hence $\phi(t)$ cannot be positive or negative in any subinterval and must therefore vanish at all points. Hence the optimal trajectory must satisfy, at all points,

$$\phi(t) = \frac{\partial F}{\partial X} - \frac{d}{dt}\frac{\partial F}{\partial \dot{X}} = 0 \tag{11.6}$$

This equation is called the *Euler-Lagrange equation*.

11.1.2 Transversality conditions

If the Euler-Lagrange equation is satisfied at all points, eqn. 11.4 reduces to

$$\left[\left\{(\dot{g}-\dot{X})\frac{\partial F}{\partial \dot{X}}+F\right\}\Delta T\right]_{T_0}^{T_1} = 0$$

If we now consider a further class of variations which go through (T_0, X_0) but *not* through (T_1, X_1), or *vice versa*, it follows that

$$\Delta T\left\{(\dot{g}-\dot{X})\frac{\partial F}{\partial \dot{X}}+F\right\} = 0 \text{ at each end-point} \qquad (11.7a)$$

Since moreover $\dot{g} = \Delta X/\Delta T$, this may also be written

$$\left(F-\dot{X}\frac{\partial F}{\partial \dot{X}}\right)\Delta T+\frac{\partial F}{\partial \dot{X}}\Delta X = 0 \text{ at each end-point} \qquad (11.7b)$$

In either form, these end-point relations are called the *transversality conditions*. Thus the optimal trajectory must satisfy both the Euler-Lagrange equation and the transversality conditions: the former, when integrated, gives the general form of the trajectory, including however integration constants, these constants being determinable by the latter. The intersections of the optimal trajectory with the boundary curves $x = g_1(t)$, $x = g_2(t)$ then determine the co-ordinates of the end-points. It must be noted, however, that the criteria developed above apply to an optimal trajectory, whether it maximises or minimises the value of I_x; if there are multiple solutions, one is still faced with the problem of whether the 'optimal' trajectory is a minimum or a maximum, possibly also with the problem of selecting the minimum minimum or the maximum maximum. One way to solve both these problems is of course to evaluate I_x along each of the optimal trajectories found; an alternative method of differentiating between maxima and minima lies in the investigation of second-order terms but this method is usually laborious. In most simple problems, the nature of the turning value(s) is fairly easily determined from the nature of the problem. Before presenting an example, however, we consider some important particular cases.

11.1.3 Important particular cases

(i) In those cases of frequent occurrence where F is not *explicitly* a function of time, i.e. $F = F(x, \dot{x})$, a first integral of the Euler-Lagrange equation is readily obtainable. For in such cases

$$\frac{dF}{dt} = \frac{\partial F}{\partial X}\dot{X} + \frac{\partial F}{\partial \dot{X}}\ddot{X}$$

Also

$$\frac{d}{dt}\left\{\dot{X}\frac{\partial F}{\partial \dot{X}}\right\} = \ddot{X}\frac{\partial F}{\partial \dot{X}} + \dot{X}\frac{d}{dt}\frac{\partial F}{\partial \dot{X}}$$

Hence, by subtraction,

$$\frac{d}{dt}\left\{\dot{X}\frac{\partial F}{\partial \dot{X}} - F\right\} = \dot{X}\left\{\frac{d}{dt}\frac{\partial F}{\partial \dot{X}} - \frac{\partial F}{\partial X}\right\}$$

$$= 0 \text{ by eqn. 11.6}$$

Hence

$$\dot{X}\frac{\partial F}{\partial \dot{X}} - F = \text{constant} \tag{11.8}$$

is a first integral of the Euler-Lagrange equation and provides a first order differential equation for $X(t)$. If this equation can be integrated, this process introduces a second constant, both constants being adjusted to satisfy the transversality conditions.

(ii) If the limits $t = t_0$, $t = t_1$, of the integral I_x are fixed, i.e. if the boundary curves $x = g_0(t)$, $x = g_1(t)$ degenerate in Fig. 11.1 to the vertical lines $t = t_0 = T_0$, $t = t_1 = T_1$ for all x, then, in eqn. 11.7*b*, $\Delta T = 0$ and it follows that at each end-point

$$\frac{\partial F}{\partial \dot{X}} = 0 \tag{11.9}$$

If only one end-value of t is fixed, then clearly eqn. 11.9 applies to that particular end-point but the more general conditions eqn. 11.7 apply to the other.

(iii) If I_x is evaluated not merely between fixed values of t, as in (ii), but also between fixed values of x, i.e. between fixed points in the (t, x) plane, then $\Delta T = \Delta X = 0$ at each end-point and the transversality conditions of eqn. 11.7 are automatically satisfied; to take their place, however, the optimal trajectory is now constrained to go through two given points.

(iv) The optimal trajectory $x = X(t)$ does *not* in general cut the boundary curves $x = g_0(t)$, $x = g_1(t)$, orthogonally. *If* it does, then $\dot{X}\dot{g} = -1$ at each end-point; hence, substituting for $\dot{g}$ in eqn. 11.7*a*, assuming $\Delta T \neq 0$:

$$\frac{\partial F}{\partial \dot{X}} = \frac{F}{\dot{X} - \dot{g}} = \frac{\dot{X}F}{1 + \dot{X}^2} \quad \text{or} \quad \frac{\partial \log F}{\partial \log (1 + \dot{X}^2)^{1/2}} = 1$$

Hence, integrating partially,

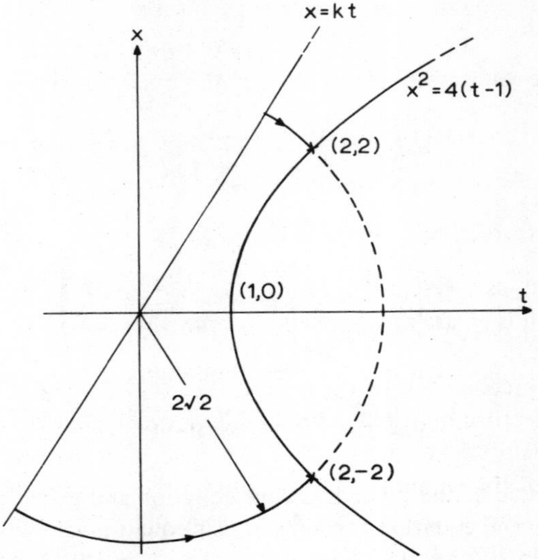

Fig. 11.2 Optimal trajectories

$$\begin{aligned}\log F &= \log(1+\dot{X}^2)^{1/2} + \text{some function of } X \text{ and } t \text{ only}\\ &= \log(1+\dot{X}^2)^{1/2} + \log h(X,t) \text{ say, so that}\\ F &= (1+\dot{X}^2)^{1/2}.h(X,t)\end{aligned}$$

Thus if F is of this form, the optimal trajectory cuts the boundaries orthogonally. Note as a particular case that, if $h = 1$, $F.dt = \{(dt)^2 + (dX)^2\}^{1/2} = ds$, an element of arc-length of the trajectory: the problem here is therefore to find the shortest path between the boundaries, the solution being a straight line cutting both boundaries at right angles.

11.1.4 Example

It is required to determine a trajectory $x(t)$ which minimises $I_x = \int \frac{1}{x}(1+\dot{x}^2)^{1/2}dt$ along a path between the straight line $x = kt$, $|k| > 1$, and the parabola $x^2 = 4(t-1)$. (Note: it is left to the reader to verify that the condition $|k| > 1$ prevents the straight line from cutting the parabola.)

Solution. Refer to Fig. 11.2. Since F, the integrand of I_x, does not contain t explicitly, a first integral of the Euler-Lagrange equation is given by eqn. 11.8: $\dot{X}\dfrac{\partial F}{\partial \dot{X}} - F = \text{constant} = \dfrac{\dot{X}^2}{X(1+\dot{X}^2)^{1/2}} - \dfrac{1}{X}(1+\dot{X}^2)^{1/2}$ in this case, i.e. on simplifying, $X(1+\dot{X}^2)^{1/2} = \text{constant} = a$, say. Then

$$\dot{X} = \frac{1}{X}(a^2 - X^2)^{1/2} \quad \text{or} \quad dt = \frac{XdX}{(a^2 - X^2)^{1/2}} \quad \text{or} \quad t = b - (a^2 - X^2)^{1/2}$$

i.e.

$$X^2 + (t-b)^2 = a^2$$

showing that, as a deduction from the Euler-Lagrange equation, the optimal path is an arc of a circle of radius a with centre at $(b, 0)$ on the t-axis.

Since, moreover, F is of the form $(1+\dot{x}^2)^{1/2}$.(function of x), the optimal trajectory must cut both boundaries orthogonally. To cut $x = kt$ orthogonally, the centre of the circle must be at the origin, so that $b = 0$. To cut the parabola orthogonally requires that at the points of intersection, the tangent to the parabola should be a radius of the circle, namely a line through the origin. Any line $x = mt$ through the origin cuts the parabola where $m^2t^2 - 4t + 4 = 0$ and therefore touches the parabola if this equation has equal roots, namely if $m^2 = 1$, and then $(t-2)^2 = 0$. Hence $t = 2$ and $x = \pm 2$ at the points of intersection of circle and parabola; hence $a^2 = 8$ and the optimal trajectory is an arc of the circle $x^2 + t^2 = 8$, giving the two trajectories shown in the diagram.

It will be readily appreciated that if the path of integration is from the origin to the vertex of the parabola, following the t-axis, then the value of $I_x \to \pm\infty$ according to whether we travel just above or just below the t-axis. It is also evident that if we follow a horizontal path ($\dot{x} = 0$) along $x = c$, say, then $I_x \to \pm\infty$ as $c \to \pm\infty$. It is clear therefore that the circular path in the upper path of the diagram *min*imises I_x while that in the lower part of the diagram *max*imises I_x.

The calculation of I_x along each of the circular arcs is left as a straightforward exercise in integration for the reader.

11.1.5 A word of warning

The example just solved, like the overwhelming majority of textbook examples, is too simple to be typical. This example is simplified by a number of factors: (*a*) a first integral of the Euler-Lagrange equation is at once obtainable because F is not explicitly a function of t; nevertheless this situation occurs fairly frequently; (*b*) the resulting first-order differential equation for X was *analytically* soluble, whereas most nonlinear equations, even of the first order, are not; (*c*) the transversality conditions are simplified to an orthogonality condition, which, in view of (*b*) and the simple geometry of the bounding curves, is easy to satisfy. In a more general case, the Euler-Lagrange equation would probably have to be solved by numerical, step-by-step methods, on a computer; a point which needs stressing in this connection is that at whichever end of the trajectory we begin the solution process, the boundary conditions (transversality conditions) at that end are insufficient to make the solution unique, since the solution has to satisfy further boundary conditions at the other end of the trajectory. The process of solution therefore necessarily involves trial and error, varying the initial conditions at the starting end until the terminal conditions are satisfied. We are in fact confronted with what is usually termed, legitimately, the *two-point boundary problem*. This problem will frequently occur in practice and is only simply overcome in those cases where an analytical solution is possible.

11.2 Extension of problem to n-dimensional space

We shall next consider an extension of the original problem to the case where $x(t)$, instead of being a scalar quantity, is an n-vector, $\boldsymbol{x}(t)$, with elements $x_1(t), x_2(t), \ldots, x_n(t)$. While thereby complicating the problem we shall at the same time simplify it somewhat by assuming that the limits of the integral, $t = t_0$ and $t = t_1$, are fixed. Under these conditions it is unnecessary to consider time as a separate dimension and we may consider trajectories in the n-dimensional space of $\boldsymbol{x}(t)$, starting from $\boldsymbol{x}(t) = \boldsymbol{x}_0$ and terminating in $\boldsymbol{x}(t) = \boldsymbol{x}_1$, where, however, $\boldsymbol{x}_0$ and $\boldsymbol{x}_1$ are in general adjustable and not given. The modified problem becomes:

Problem II: Find $\boldsymbol{x}(t)$ such that $I_{\boldsymbol{x}} = \int_{t_0}^{t_1} F(\boldsymbol{x}, \dot{\boldsymbol{x}}, t)\,dt$ shall have a turning value, t_0 and t_1 being given, $\boldsymbol{x}(t_0) \equiv \boldsymbol{x}_0$ and $\boldsymbol{x}(t_1) \equiv \boldsymbol{x}_1$ being either arbitrary or, as a particular case, given.

By $F(\boldsymbol{x}, \dot{\boldsymbol{x}}, t)$ a scalar function *of the elements of* $\boldsymbol{x}(t)$ and $\dot{\boldsymbol{x}}(t)$ and time t is implied, which is assumed differentiable with respect to these $(2n + 1)$ arguments.

Solution. We assume, as in the simpler problem, that there is a solution to the problem, namely that there is some $\boldsymbol{x}(t) = \boldsymbol{X}(t)$ such that $I_{\boldsymbol{X}}$ is a turning value of $I_{\boldsymbol{x}}$, this trajectory having end-points at $\boldsymbol{x} = \boldsymbol{X}(t_0) = \boldsymbol{X}_0$ and $\boldsymbol{x} = \boldsymbol{X}(t_1) = \boldsymbol{X}_1$. We consider, as before, a neighbouring trajectory

$x(t) = X(t) + \delta X(t)$ with end-points at $X(t_0) + \delta X(t_0) \equiv X_0 + \delta X_0$ and $X(t_1) + \delta X(t_1) \equiv X_1 + \delta X_1$. The notation used implies that at every t, each element of $\delta X(t)$ is a vanishingly small quantity. The increment in I_x caused by moving from the optimal trajectory to its neighbour is

$$\delta I_X = I_{X+\delta X} - I_X = \int_{t_0}^{t_1} \{F(X + \delta X, \dot{X} + \delta\dot{X}, t) - F(X, \dot{X}, t)\}\, dt$$

Assuming as before that F is differentiable with respect to all its arguments, we may expand the first term of the integrand as a Taylor series, retaining only first-order terms:

$$F(X + \delta X, \dot{X} + \delta\dot{X}, t)$$
$$= F(X, \dot{X}, t) + \frac{\partial F}{\partial X}\delta X + \frac{\partial F}{\partial \dot{X}}\delta\dot{X} + \text{higher order terms.}$$

(Note that in this expression the partial derivative of the *scalar* F with respect to the *column* X is defined as the *row* $\left\{\frac{\partial F}{\partial X_1}, \frac{\partial F}{\partial X_2}, \ldots, \frac{\partial F}{\partial X_n}\right\}$. This is a matter of convention and some authors prefer to define this derivative as the *column* of these elements.) Hence, to a first order of small quantities,

$$\delta I_X = \int_{t_0}^{t_1} \left\{\frac{\partial F}{\partial X}\delta X + \frac{\partial F}{\partial \dot{X}}\delta\dot{X}\right\} dt = 0$$

But, integrating by parts,

$$\int_{t_0}^{t_1} \frac{\partial F}{\partial \dot{X}}\delta\dot{X}.dt = \left[\frac{\partial F}{\partial \dot{X}}\delta X\right]_{t_0}^{t_1} - \int_{t_0}^{t_1} \frac{d}{dt}\frac{\partial F}{\partial \dot{X}}.\delta X.dt$$

Hence

$$\delta I_X = \int_{t_0}^{t_1} \left\{\frac{\partial F}{\partial X} - \frac{d}{dt}\frac{\partial F}{\partial \dot{X}}\right\} \delta X.dt + \left[\frac{\partial F}{\partial \dot{X}}\delta X\right]_{t_0}^{t_1} = 0 \qquad (11.10)$$

for all differentiable δX.

As in Problem I we first consider variations δX which vanish at both end-points. For such variations, eqn. 11.10 reduces to

$$\int_{t_0}^{t_1} \left\{\frac{\partial F}{\partial X} - \frac{d}{dt}\frac{\partial F}{\partial \dot{X}}\right\} \delta X.dt = 0$$

which is clearly satisfied by making

$$\frac{\partial F}{\partial X} = \frac{d}{dt}\frac{\partial F}{\partial \dot{X}} \qquad (11.11)$$

It may be shown, by an extension of the method used in the preliminary problem, that this condition is not only sufficient but necessary if the previous integral is to vanish for *all* variations of the type considered. This condition, eqn. 11.11, is the multidimensional form of the Euler-Lagrange equation: it represents, of course, n scalar differential equations.

If this vector equation is satisfied, eqn. 11.10 reduces to $\left[\frac{\partial F}{\partial \dot{X}} \delta X\right]_{t_0}^{t_1} = 0$. By choosing variations such that δX vanishes at $t = t_0$ but not at $t = t_1$ or *vice versa*, we deduce that at *each* end-point of the optimal trajectory

$$\frac{\partial F}{\partial \dot{X}} \delta X = 0 \tag{11.12}$$

which is the simple form taken in this problem by the transversality conditions, simple because t_0 and t_1 have been assumed fixed. By considering variations such that at either end-point $\delta X_r(t) \neq 0$ but $\delta X_s(t) = 0$ ($s = 1, 2, \ldots, n$ excluding r), it is at once deducible from eqn. 11.12 that at each end-point

$$\frac{\partial F}{\partial \dot{X}_r} \delta X_r = 0 \qquad (r = 1, 2, \ldots, n) \tag{11.12a}$$

It follows that if, at either $t = t_0$ or $t = t_1$, the value of $x_r(t)$ is specified, then at that end-point $X_r(t)$ must have this specified value, with the result that $\delta X_r(t)$ must vanish at that end-point, thus satisfying eqn. 11.12a automatically; the optimal trajectory is however constrained by the dictated value of $X_r(t)$ at that end-point. If however $x_r(t)$ is *not* specified at one or other end-point, then $\delta X_r(t)$ may have any small value and to satisfy eqn. 11.12a now requires $\frac{\partial F}{\partial \dot{X}_r} = 0$ at that end-point.

Finally, as in Problem I, if F is not explicitly a function of t, a first integral of the Euler-Lagrange equation is always obtainable as

$$\frac{\partial F}{\partial \dot{X}} \dot{X} - F = \text{constant} \tag{11.13}$$

The proof is similar to that of Section 11.1.3 (i) and is left to the reader.

11.3 Optimisation with state equation constraint

Problem III: A first order ($n = 1$) system obeys the equation

$$\dot{x} = f(x, u, t)$$

It is required to find $u(t)$ so that the integral

$I = \int_{t_0}^{t_1} G(x, u, t)\,dt$ shall have a stationary value, t_0 and t_1 being fixed.

Solution. Note first that *if* the system equation can be solved explicitly for u in the form, say, $u = g(x, \dot{x}, t)$, then the integrand of I may be written $G\{x, g(x, \dot{x}, t), t\} \equiv F(x, \dot{x}, t)$, say, and the problem is then identical with Problem I; for if $x(t)$ can be found, so can $\dot{x}(t)$ and hence so can $u(t) = g(x, \dot{x}, t)$. We shall assume that this procedure cannot be followed.

Note further that $x(t)$ and $u(t)$ are not independent functions of t since they are related by the system equation.

Suppose a solution $u(t) = U(t)$ exists, associated with $x(t) = X(t)$, the latter having terminal values $X(t_0) = X_0$ and $X(t_1) = X_1$ which may or may not be specified. We visualise a 'neighbouring' pair of functions which, to illustrate a slightly different technique from that used previously, we shall denote by $u(t) = U(t) + \epsilon v(t)$ and $x(t) = X(t) + \epsilon y(t)$, where v and y are arbitrary but differentiable functions and ϵ is a vanishingly small real number. For simplicity we shall also use the suffix notation for partial derivatives, i.e. $\dfrac{\partial f}{\partial x} \equiv f_x$ etc. We have

$$\delta I = \int_{t_0}^{t_1} \{G(X + \epsilon y, U + \epsilon v, t) - G(X, U, t)\}\,dt$$

$$= \int_{t_0}^{t_1} \epsilon(yG_X + vG_U)\,dt \text{ (to a first order in } \epsilon) = 0$$

or, since ϵ, though small, is not zero, we deduce

$$\int_{t_0}^{t_1} (yG_X + vG_U)\,dt = 0 \qquad (11.14)$$

But since the function pairs must each satisfy the system equation,

$$\dot{X} = f(X, U, t) \quad \text{and} \quad \dot{X} + \epsilon\dot{y} = f(X + \epsilon y, U + \epsilon v, t)$$

By subtraction and the use of a Taylor series, we obtain

$$\dot{y} = yf_X + vf_U \quad \text{or} \quad v = (\dot{y} - yf_X)/f_U$$

Substituting this value in eqn. 11.14 gives

$$\int_{t_0}^{t_1} \{\dot{y}G_U/f_U + y(G_X - G_Uf_X/f_U)\}\,dt = 0$$

Integrating the first term by parts gives

$$\int_{t_0}^{t_1} \dot{y}.(G_U/f_U)\,dt = [yG_U/f_U]_{t_0}^{t_1} - \int_{t_0}^{t_1} y\frac{d}{dt}(G_U/f_U)\,dt$$

which, substituted in the previous equation, gives

$$\int_{t_0}^{t_1} y\left\{G_X - G_U f_X/f_U - \frac{d}{dt}(G_U/f_U)\right\}dt + [yG_U/f_U]_{t_0}^{t_1} = 0$$

an equation which must be satisfied for *all* y-functions.

Using the same methods as in the earlier problems, we deduce as necessary and sufficient conditions:

$$G_X - G_U f_X/f_U - \frac{d}{dt}(G_U/f_U) = 0 \quad \text{(Euler-Lagrange equation)} \tag{11.15}$$

$$yG_U/f_U = 0 \text{ at } t = t_0 \text{ and } t = t_1 \quad \text{(transversality conditions)} \tag{11.16}$$

If, at either terminal time, x is specified, then $y = 0$ and eqn. 11.16 is automatically satisfied; if, at either terminal time, x is 'free', then $y \neq 0$ and eqn. 11.16 requires $G_U/f_U = 0$.

11.3.1 Use of the Lagrange multiplier

We now present, subject to verification, an alternative method of solving the same problem. Write the constraining system equation in the form $f(x, u, t) - \dot{x} = 0$ and add to the integrand of I a time-varying multiplier, $k(t)$, of the left hand side of this equation to form the *augmented integral*

$$I_a = \int_{t_0}^{t_1} [G(x, u, t) + k(t)\{f(x, u, t) - \dot{x}\}]\,dt$$

The function $k(t)$ is called a *Lagrange multiplier*.

Consider now the problem of finding stationary values of I_a *without constraints*. Its integrand is a function of three, now independent, time-functions, $x(t)$, $u(t)$ and $k(t)$, as well as of time t; note however that $\dot{u}$ and $\dot{k}$ do not appear. Considering x, u, k as the elements of a vector of order 3, the problem is therefore identical with Problem II with $n = 3$. Thus, using eqn. 11.11, the Euler-Lagrange equation for this problem, but replacing X_1, X_2, X_3 by the optimising values X, U, K of x, u, k and replacing F by the integrand of I_a, we obtain:

$$\text{from} \quad \frac{\partial F}{\partial X} = \frac{d}{dt}\frac{\partial F}{\partial \dot{X}} \quad : \quad G_X + Kf_X = -\dot{K} \tag{11.17a}$$

$$\text{from} \quad \frac{\partial F}{\partial U} = \frac{d}{dt}\frac{\partial F}{\partial \dot{U}} \quad : \quad G_U + Kf_U = 0 \tag{11.17b}$$

$$\text{from} \quad \frac{\partial F}{\partial K} = \frac{d}{dt}\frac{\partial F}{\partial \dot{K}} \quad : \quad f(X, U, t) - \dot{X} = 0 \tag{11.17c}$$

We note first that the problem of finding a stationary value of I_a demands, by eqn. 11.17*c*, that the system equation should be satisfied. Moreover, if this is the case, then I_a reduces to I. In other words, *finding stationary values of I_a without constraints is identical to finding stationary values of I in the presence of the constraining system equation.*

Moreover, from eqn. 11.17*b*, $K = -G_U/f_U$ and substituting in eqn. 11.17*a* gives precisely eqn. 11.15, the form of the Euler-Lagrange equation obtained by the first method.

As far as the transversality conditions are concerned, we revert to eqn. 11.12 of Problem II, which, similarly translated into the symbols of the present problem, gives: $[-K, 0, 0][\delta X, \delta U, \delta K]' = 0$, i.e. $K.\delta X = 0$. Since $\delta X = \epsilon y$ and, from eqn. 11.17*b*, $K = -G_U/f_U$, we deduce $yG_U/f_U = 0$ which is again the same result as that obtained in eqn. 11.16 by the first method.

The introduction of the unknown Lagrange multipliers and the formation of the augmented integral has therefore been shown to give identical results to those of the first method. This new method has the advantage that it is easily generalised to give the solution to similar problems when $n > 1$. Before demonstrating this, however, we consider an example.

11.3.2 Example

Let the first-order system equation be $\dot{x} = -ax + bu$ and let the integral to be minimised be $I = \int_0^T (x^2 + u^2)\,dt$, T fixed.

Solution: Using the Lagrange multiplier $k(t)$, we form the augmented integral $\int_0^T \{x^2 + u^2 + k(-ax + bu - \dot{x})\}\,dt$. Writing down the Euler-Lagrange equations in the form of eqn. 11.17 we obtain:

$2x - ak = -\dot{k}$; $2u + bk = 0$; $-ax + bu - \dot{x} = 0$ (the system equation)

From the second of these, $u = -bk/2$ or $k = -2u/b$. Substituting in the first gives $\dot{u} = au + bx$, which, together with the system equation, provides two simultaneous first-order differential equations for x and u. With $D \equiv d/dt$, $(D-a)u = bx$, $(D+a)x = bu$, so that on multiplying, $D^2 - a^2 = b^2$ or $D^2 = a^2 + b^2 \equiv c^2$ say, and $D = \pm c$. Postulating $x = A \cosh ct + B \sinh ct$ and using $u = (D+a)x/b$ leads to $u = \{(cB + aA) \cosh ct + (cA + aB) \sinh ct\}/b$. A and B are to be found from the transversality conditions. At $t = 0$, if x is given and equal to x_0, then $A = x_0$; if $x(0)$ is free, then at $t = 0$, $\dfrac{\partial F}{\partial \dot{x}} = -k = 0$, and since $k = -2u/b$, it follows that $cB + aA = 0$. Similarly, if, at $t = T$, $x = x_T$, given, then $A \cosh cT + B \sinh cT = x_T$; but if $x(T)$ is free, then $k(T) = u(T) = 0$, i.e. $(cB + aA) \cosh cT + (cA + aB) \sinh cT = 0$. Thus whether x is fixed or free at either end-point, we have two equations from which to determine A and B and hence $x(t)$, $u(t)$ and $k(t)$. Now read again Section 11.1.5!

11.4 Extension of method to nth order state vector

Problem IV Given a system representable by the vector equation $\dot{\boldsymbol{x}} = \boldsymbol{f}(\boldsymbol{x}, \boldsymbol{u}, t)$ where $\boldsymbol{x}$ is an n-vector and $\boldsymbol{u}$ is an m-vector $(m \leqslant n)$, determine $\boldsymbol{u}(t)$ so that the integral $I = \int_{t_0}^{t_1} G(\boldsymbol{x}, \boldsymbol{u}, t)\, dt$ has a turning value, G being differentiable with respect to all its arguments. Investigate the effect of fixed or free terminal values of $\boldsymbol{x}(t)$.

Solution: First form the augmented integral

$$I_a = \int_{t_0}^{t_1} [G(\boldsymbol{x}, \boldsymbol{u}, t) + \boldsymbol{k}'(t)\{\boldsymbol{f}(\boldsymbol{x}, \boldsymbol{u}, t) - \dot{\boldsymbol{x}}\}]\, dt \equiv \int_{t_0}^{t_1} F(\boldsymbol{x}, \boldsymbol{u}, \boldsymbol{k}, \dot{\boldsymbol{x}}, t)\, dt$$

in which $\boldsymbol{k}(t)$ is a column of n elements. In Problem II, consider $\boldsymbol{X}(t)$ to be a column of $(2n+m)$ elements, namely $\boldsymbol{X} = [\boldsymbol{x}', \boldsymbol{u}', \boldsymbol{k}']'$. Then the Euler-Lagrange equation, eqn. 11.11 may be broken up into

$$\frac{\partial F}{\partial \boldsymbol{x}} = \frac{d}{dt}\frac{\partial F}{\partial \dot{\boldsymbol{x}}} \quad \text{i.e.} \quad \frac{\partial G}{\partial \boldsymbol{x}} + \boldsymbol{k}'J(\boldsymbol{f}:\boldsymbol{x}) = -\dot{\boldsymbol{k}}' \qquad (11.18a)$$

$$\frac{\partial F}{\partial \boldsymbol{u}} = \frac{d}{dt}\frac{\partial F}{\partial \dot{\boldsymbol{u}}} \quad \text{i.e.} \quad \frac{\partial G}{\partial \boldsymbol{u}} + \boldsymbol{k}'J(\boldsymbol{f}:\boldsymbol{u}) = \boldsymbol{0} \qquad (11.18b)$$

$$\frac{\partial F}{\partial \boldsymbol{k}} = \frac{d}{dt}\frac{\partial F}{\partial \dot{\boldsymbol{k}}} \quad \text{i.e.} \quad \boldsymbol{f}'(\boldsymbol{x}, \boldsymbol{u}, t) - \dot{\boldsymbol{x}}' = \boldsymbol{0} \qquad (11.18c)$$

in which $J(\boldsymbol{f}:\boldsymbol{x})$ denotes the Jacobian matrix of the column $\boldsymbol{f}$ with respect to the column x, having the typical element $j_{rs} = \dfrac{\partial f_r}{\partial x_s}$, a similar definition applying to $J(f:u)$. The third equation, eqn. 11.18c shows that the optimal trajectory must satisfy the state equation, in which case $I_a = I$, so that, as in Problem III, optimising I_a without constraints is identical with optimising I in the presence of the constraining state equation.

The eqns. 11.18 represent a total of $(2n + m)$ scalar equations of which, theoretically at any rate, we may consume m by the process of eliminating the m elements of u. There will remain $2n$ differential equations involving the $2n$ elements of x and k. The $2n$ boundary conditions for these equations are obtainable from the transversality conditions at $t = t_0$ and $t = t_1$, n at each terminal point. These are: at both $t = t_0$ and $t = t_1$, either an element x_r of x is given or it is free, in which case $\dfrac{\partial F}{\partial \dot{x}_r} = -k_r = 0$. Thus at each terminal point either x_r or k_r is known $(r = 1, 2, \ldots, n)$.

Although the data, whatever its nature, is therefore theoretically sufficient to solve the problem (not necessarily uniquely) the actual process of solution may be very tedious, due largely, once again, to the 2-point boundary problem: whichever end of the trajectory we start from, assuming that u has been eliminated, we have to solve for $2n$ unknowns but we only have n boundary conditions. We therefore begin the process of a step-by-step, computational solution by necessarily allocating to the n unknown quantities arbitrary values; at the other terminal point the results are to be compared with the n further boundary conditions. This leads to a trial and error process of adjusting the initial arbitrary values until the final boundary conditions are reasonably closely satisfied. There are techniques for accelerating this trial and error procedure, but we only wish to draw attention yet again to the magnitude of the problem.

11.4.1 Linear system with quadratic form criterion integrand

Consider a particular case of Problem IV:

Problem V: Given a system representable by the state equation $\dot{x} = Ax + Bu$ determine $u(t)$ so that $I = \int_{t_0}^{t_1} \frac{1}{2}(x'Px + u'Qu)\,dt$ shall be minimised, assuming such a $u(t)$ exists.

A, B, P, Q may or may not be time-dependent; P and Q may be assumed symmetric, Q positive-definite ($u'Qu > 0$ for all $u \neq 0$) and P non-negative-definite ($x'Px \geqslant 0$ for all $x \neq 0$).

Solution: The integrand of the augmented integral is in this case

$$F(x, u, k, \dot{x}, t) = \tfrac{1}{2}(x'Px + u'Qu) + k'(Ax + Bu - \dot{x})$$

Applying eqns. 11.18*a* and *b*, gives:

$$x'P + k'A = -\dot{k}' \quad \text{or, transposing,} \quad Px + A'k = -\dot{k} \qquad (11.19a)$$

$$u'Q + k'B = 0 \quad \text{or, transposing,} \quad Qu + B'k = 0 \qquad (11.19b)$$

Since Q is positive-definite and therefore nonsingular, this last equation may be solved for u to give

$$u = -Q^{-1}B'k \qquad (11.19c)$$

Substituting this value in the state equation then gives

$$\dot{x} = Ax - BQ^{-1}B'k \equiv Ax - Sk$$

where

$$S = BQ^{-1}B'$$

and is clearly symmetric. Combining this with eqn. 11.19*a* we have

$$\dot{x} = Ax - Sk \qquad (11.20a)$$

$$\dot{k} = -Px - A'k \qquad (11.20b)$$

The similarity of form of these two equations for x and k has led to the Lagrange multiplier $k(t)$ being termed the *co-state vector*; eqn. 11.20*b* is usually referred to as the *co-state equation*. The remaining equation, eqn. 11.19*b*, which we have used to eliminate u is called the *control equation*.

If any of the matrices A, B, P, Q are time-dependent (and nothing in the above analysis assumes that they are not) the solution of eqn. 11.20 is in general *analytically* impossible: in such a case the only slight advantage resulting from the assumption of a linear system is that these equations are also linear, but we still have the 2-point boundary problem to complicate any computational process of solution. If, however, A, B, P, Q, are independent of time, an analytical solution is always possible and the 2-point boundary problem is removed, as will now be shown.

11.4.2 Time-independent matrices

Write the two equations of eqn. 11.20 as a single equation:

$$\begin{bmatrix} \dot{x} \\ \dot{k} \end{bmatrix} = \begin{bmatrix} A & -S \\ -P & -A' \end{bmatrix} \begin{bmatrix} x \\ k \end{bmatrix} \equiv W \begin{bmatrix} x \\ k \end{bmatrix} \tag{11.21}$$

where W is square, of order $2n$, and may be partitioned into the four square matrices of order n shown. Suppose for simplicity that the eigenvalues of W are distinct and that an associated eigenvector assembly matrix is E.

Then the solution of eqn. 11.21 is (taking $t_0 = 0$, $t_1 = T$ for convenience)

$$\begin{bmatrix} x(t) \\ k(t) \end{bmatrix} = \exp Wt . \begin{bmatrix} x(0) \\ k(0) \end{bmatrix} = E . \operatorname{diag} \exp(\lambda_r t) . E^{-1} . \begin{bmatrix} x(0) \\ k(0) \end{bmatrix}$$

$$\equiv E . \operatorname{diag} \exp(\lambda_r t) . q \tag{11.22}$$

where the λ_r, $r = 1, 2, \ldots, 2n$, are the eigenvalues of W and where $q_{2n \times 1}$ is some constant column, the elements of which are to be determined from the boundary conditions. But at $t = 0$ and at $t = T$ either an element x_r of x is given or else it is free, in which case $k_r = 0$ ($r = 1, 2, \ldots, n$) and, from eqn. 11.22, every element of x and of k is a linear combination of the $2n$ elements of q. Inserting the boundary conditions, which, whatever their nature, are $2n$ in number, therefore gives $2n$ linear equations for these elements, the values of which may therefore be found uniquely and then substituted in eqn. 11.22 to give an explicit solution for $x(t)$ and $k(t)$. The solution of the problem is completed by deriving $u(t)$ from eqn. 11.19*b*.

11.4.2.1 Notes

(i) Due to the nature of its partitioned structure (in which the leading diagonal quadrants are, mutually, negative transposes, the other quadrants being symmetric matrices) W is known as a *Hamiltonian matrix*. Such matrices have the property that their eigenvalues always occur in equal and opposite pairs. Indeed the characteristic function of W is given by

$$w(\lambda) = \det(W - \lambda I_{2n})$$

$$= \det \begin{bmatrix} A - \lambda I_n & -S \\ -P & -A' - \lambda I_n \end{bmatrix}$$

$$= \det \begin{bmatrix} A' - \lambda I_n & -P \\ -S & -A - \lambda I_n \end{bmatrix} \text{ after transposing, remembering that } P \text{ and } S \text{ are symmetric,}$$

$$= \det \begin{bmatrix} -A - \lambda I_n & -S \\ -P & A' - \lambda I_n \end{bmatrix} \text{ after interchanging top and bottom halves and left and right halves,}$$

$$= \det \begin{bmatrix} A + \lambda I_n & -S \\ -P & -A' + \lambda I_n \end{bmatrix} \text{ after multiplying the first } n \text{ columns and the last } n \text{ rows by } (-1),$$

$$= w(-\lambda),$$

showing that if λ is an eigenvalue of $\boldsymbol{W}$, so is $(-\lambda)$. It may be convenient, therefore, to order the eigenvalues of $\boldsymbol{W}$ as $(\lambda_1, \lambda_2, \ldots, \lambda_n, -\lambda_1, -\lambda_2, \ldots, -\lambda_n)$ where, on the supposition that no eigenvalues are purely imaginary, we may suppose the first n eigenvalues to have positive real parts, the second set therefore having negative real parts. Then we may write

$$\operatorname{diag} \lambda_r, (r = 1, 2, \ldots, 2n) = \begin{bmatrix} \boldsymbol{\Delta} & 0 \\ 0 & -\boldsymbol{\Delta} \end{bmatrix}$$

where

$$\boldsymbol{\Delta} = \operatorname{diag} \lambda_r (r = 1, 2, \ldots, n)$$

$$\operatorname{diag} \exp \lambda_r t, (r = 1, 2, \ldots, 2n) = \begin{bmatrix} \boldsymbol{M}(t) & 0 \\ 0 & \boldsymbol{M}^{-1}(t) \end{bmatrix},$$

where

$$\boldsymbol{M}(t) = \operatorname{diag} \exp \lambda_r t, (r = 1, 2, \ldots, n).$$

If we further partition $\boldsymbol{E}$ as $\begin{bmatrix} \boldsymbol{E}_1 & \boldsymbol{E}_2 \\ \boldsymbol{E}_3 & \boldsymbol{E}_4 \end{bmatrix}$ and $\boldsymbol{q}$ as $\begin{bmatrix} q_1 \\ q_2 \end{bmatrix}$, the general solution to eqn. 11.22 becomes

$$\begin{bmatrix} \boldsymbol{x}(t) \\ \boldsymbol{k}(t) \end{bmatrix} = \begin{bmatrix} \boldsymbol{E}_1 & \boldsymbol{E}_2 \\ \boldsymbol{E}_3 & \boldsymbol{E}_4 \end{bmatrix} \begin{bmatrix} \boldsymbol{M}(t) & 0 \\ 0 & \boldsymbol{M}^{-1}(t) \end{bmatrix} \begin{bmatrix} q_1 \\ q_2 \end{bmatrix} = \begin{bmatrix} \boldsymbol{E}_1 \boldsymbol{M}(t) \boldsymbol{q}_1 + \boldsymbol{E}_2 \boldsymbol{M}^{-1}(t) \boldsymbol{q}_2 \\ \boldsymbol{E}_3 \boldsymbol{M}(t) \boldsymbol{q}_1 + \boldsymbol{E}_4 \boldsymbol{M}^{-1}(t) \boldsymbol{q}_2 \end{bmatrix} \tag{11.23}$$

It should be noted that if, in a particular problem, $T \to \infty$, then I is a divergent integral unless both $\boldsymbol{x}(t)$ and $\boldsymbol{u}(t) \to 0$ as $t \to \infty$, and the same must be true of $\boldsymbol{k}(t)$ due to eqn. 11.19*c*. Hence if I is to be minimised, the solutions for $\boldsymbol{x}(t)$ and $\boldsymbol{k}(t)$ can only contain positively attenuated modes, namely those associated with eigenvalues of $\boldsymbol{W}$ having negative real parts. Since $\boldsymbol{M}$, as stated above, contains only *negatively* attenuated modes, it follows that the coefficient of $\boldsymbol{M}$ in the solution for $\boldsymbol{x}$ and $\boldsymbol{k}$ must vanish, i.e. $\boldsymbol{q}_1$ must be zero. Hence in such a case eqn. 11.23 reduces to

$$\begin{bmatrix} \boldsymbol{x}(t) \\ \boldsymbol{k}(t) \end{bmatrix} = \begin{bmatrix} \boldsymbol{E}_2 \\ \boldsymbol{E}_4 \end{bmatrix} \boldsymbol{M}^{-1}(t) \boldsymbol{q}_2$$

the elements of $\boldsymbol{q}_2$ being obtainable from the n boundary conditions of the problem at $t = 0$.

(ii) Alternatively, if the matrix exp $\boldsymbol{W}t$ can be computed conveniently, we may revert to the first form of the solution of eqn. 11.22, namely

$$\begin{bmatrix} \boldsymbol{x}(t) \\ \boldsymbol{k}(t) \end{bmatrix} = \exp \boldsymbol{W}t \begin{bmatrix} \boldsymbol{x}(0) \\ \boldsymbol{k}(0) \end{bmatrix} = \exp \boldsymbol{W}(t-T) \begin{bmatrix} \boldsymbol{x}(T) \\ \boldsymbol{k}(T) \end{bmatrix}$$

Partitioning $\exp \boldsymbol{W}t \equiv \begin{bmatrix} \boldsymbol{L}_1(t) & \boldsymbol{L}_2(t) \\ \boldsymbol{L}_3(t) & \boldsymbol{L}_4(t) \end{bmatrix}$, leads to

$$\begin{bmatrix} \boldsymbol{x}(t) \\ \boldsymbol{k}(t) \end{bmatrix} = \begin{bmatrix} \boldsymbol{L}_1(t).\boldsymbol{x}(0) + \boldsymbol{L}_2(t).\boldsymbol{k}(0) \\ \boldsymbol{L}_3(t).\boldsymbol{x}(0) + \boldsymbol{L}_4(t).\boldsymbol{k}(0) \end{bmatrix} = \begin{bmatrix} \boldsymbol{L}_1(t-T).\boldsymbol{x}(T) + \boldsymbol{L}_2(t-T).\boldsymbol{k}(T) \\ \boldsymbol{L}_3(t-T).\boldsymbol{x}(T) + \boldsymbol{L}_4(t-T).\boldsymbol{k}(T) \end{bmatrix} \tag{11.24}$$

In this form of the solution there are $4n$ scalar equations from which the $2n$ *unknown* elements of $\boldsymbol{x}(0), \boldsymbol{x}(T), \boldsymbol{k}(0), \boldsymbol{k}(T)$, (the other $2n$ being known from the boundary conditions) may be eliminated, leaving $2n$ equations for $\boldsymbol{x}(t)$ and $\boldsymbol{k}(t)$.

11.4.3 The Riccati matrix

This Chapter is concluded by the introduction of an alternative method of solving the problem of Section 11.4.1, applicable whether $\boldsymbol{A}, \boldsymbol{B}, \boldsymbol{P}, \boldsymbol{Q}$ are time-independent or not.

Suppose that a matrix $\boldsymbol{N}(t)$ may be found such that

$$\boldsymbol{k}(t) = \boldsymbol{N}(t).\boldsymbol{x}(t) \tag{11.25}$$

Viewed in isolation, this presents no difficulty, for since $k_r(t) = \sum_{s=1}^{n} n_{rs}(t)x_s(t)$ the n elements of any row of $\boldsymbol{N}$ only have to satisfy a single equation, so that there are an infinite number of solutions for $\boldsymbol{N}(t)$ (unless $n = 1$). Substituting for $\boldsymbol{k}(t)$ in eqn. 11.20 now leads to

$$\dot{\boldsymbol{x}} = (\boldsymbol{A} - \boldsymbol{SN})\boldsymbol{x} \quad \text{and} \quad \dot{\boldsymbol{N}}\boldsymbol{x} + \boldsymbol{N}\dot{\boldsymbol{x}} = -(\boldsymbol{P} + \boldsymbol{A}'\boldsymbol{N})\boldsymbol{x}$$

Substituting for $\dot{\boldsymbol{x}}$ from the first of these equations in the second gives

$$(\dot{\boldsymbol{N}} + \boldsymbol{NA} + \boldsymbol{A}'\boldsymbol{N} + \boldsymbol{P} - \boldsymbol{NSN})\boldsymbol{x} = \boldsymbol{0}$$

and if this equation is to be satisfied by *all* optimising trajectories, irrespective of the boundary conditions of the problem, then

$$\dot{\boldsymbol{N}} + \boldsymbol{NA} + \boldsymbol{A}'\boldsymbol{N} + \boldsymbol{P} - \boldsymbol{NSN} = \boldsymbol{0} \tag{11.26}$$

This first-order, nonlinear differential equation for $\boldsymbol{N}(t)$ is said to be of the Riccati type and the matrix N is called the Riccati matrix for the problem. If $N(t)$ can be found, then $x(t)$ may be derived from $\dot{x} = (A - SN)x$ (given above), $k(t)$ from $k = Nx$ and $u(t)$ from eqn. 11.19*c*, $u = -Q^{-1}B'k$. The differential equations for N and for x of course require initial conditions for their solution, which is necessarily a computerised process; but the point we wish to stress here is that if N can be found, then $u = -Q^{-1}B'k = -Q^{-1}B'N.x$. In other words *the input $u(t)$ required to minimise the integral I of Problem V may be obtained as a feed-back from the state-vector*, this feedback being, however, time-dependent in general.

As we shall be returning to the Riccati matrix in Chapter 13, we shall defer discussing the major problem of the boundary conditions applying to eqn. 11.26 until then. We conclude with some important properties of the Riccati matrix $N(t)$.

11.4.3.1 Some properties of the Riccati matrix

(i) If we transpose eqn. 11.26 term by term, interchanging the second and third terms, we obtain

$$\dot{N}' + N'A + A'N' + P - N'SN' = 0$$

showing that N' satisfies the same Riccati equation as N.

(ii) Since $NN^{-1} = I_n$, it follows that $\dot{N}N^{-1} + N\frac{d}{dt}N^{-1} = 0$ or $\dot{N} = -N\frac{dN^{-1}}{dt}N$. On substituting this value in eqn. 11.26 and then pre- and post-multiplying by N^{-1}, we deduce

$$\frac{dN^{-1}}{dt} - AN^{-1} - N^{-1}A' + S - N^{-1}PN^{-1} = 0$$

which shows that N^{-1} satisfies a 'dual' Riccati equation, obtainable from eqn. 11.26 by changing A into $-A'$ (and therefore A' into $-A$) and interchanging P and S. This is to be expected, for if $k = Nx$, then $x = N^{-1}k$, and it follows that changing N to N^{-1} is equivalent to interchanging x and k in eqn. 11.20.

(iii) *Twice* the integrand of I in Problem V is equal to

$$\begin{aligned} x'Px + u'Qu &= x'Px + x'N'BQ^{-1}.Q.Q^{-1}B'Nx \\ &\qquad\qquad (\text{since } u = -Q^{-1}B'Nx) \\ &= x'(P + N'SN)x \qquad (\text{since } S = BQ^{-1}B') \\ &= x'(NSN - NA - A'N + N'SN - \dot{N})x, \text{ (on substituting for } P \text{ from eqn. 11.26)} \end{aligned}$$

$$= \boldsymbol{x}'\{(N+N')SN-(N+N')A-\dot{N}\}\boldsymbol{x},$$

(for $\boldsymbol{x}'A'N\boldsymbol{x}$ = scalar = its transpose $= \boldsymbol{x}'N'A\boldsymbol{x}$)

$$= -\boldsymbol{x}'(N+N')\dot{\boldsymbol{x}}-\boldsymbol{x}'\dot{N}\boldsymbol{x},$$

since $\dot{\boldsymbol{x}} = (A-SN)\boldsymbol{x}$,

$$= -\boldsymbol{x}'N\dot{\boldsymbol{x}}-\dot{\boldsymbol{x}}'N\boldsymbol{x}-\boldsymbol{x}'\dot{N}\boldsymbol{x},$$

since $\boldsymbol{x}'N'\dot{\boldsymbol{x}}$ = scalar $= \dot{\boldsymbol{x}}'N\boldsymbol{x}$

$$= -\frac{d}{dt}(\boldsymbol{x}'N\boldsymbol{x})$$

It therefore follows that the required minimum value of I is

$$I_{min} = \tfrac{1}{2}[\boldsymbol{x}'N\boldsymbol{x}]_{t_1}^{t_0} \tag{11.27}$$

It also follows that, provided the optimising value of $\boldsymbol{u}(t)$ is used with its optimal trajectory $\boldsymbol{x}(t)$, then

$$\int_t^{t_1} \tfrac{1}{2}(\boldsymbol{x}'P\boldsymbol{x}+\boldsymbol{u}'Q\boldsymbol{u})\,dt = \tfrac{1}{2}[\boldsymbol{x}'N\boldsymbol{x}]_{t_1}^{t}$$

and

$$\int_{t_0}^{t} \tfrac{1}{2}(\boldsymbol{x}'P\boldsymbol{x}+\boldsymbol{u}'Q\boldsymbol{u})\,dt = \tfrac{1}{2}[\boldsymbol{x}'N\boldsymbol{x}]_{t}^{t_0}$$

11.5 Examples 11

1. Find the trajectories in the plane of (t, x) which will minimise

$$I = \int_0^T \frac{1+\dot{x}^2}{x}\,dx \qquad (T>0 \text{ but unspecified})$$

between the boundaries $t=0$ and $x^2 = t-1$, in each of the three cases
(i) $x(0)$ and $x(T)$ both free
(ii) $x(0)=1$ but $x(T)$ free
(iii) $x(0)=x(T)=1$.
In all cases $x(t)$ is to be positive for all t in $0 \leqslant t \leqslant T$.

2. A system is controlled by the equations

$$\dot{\boldsymbol{x}} = \begin{bmatrix} 0 & 1 \\ 0 & -5 \end{bmatrix}\boldsymbol{x} + \begin{bmatrix} 0 \\ 12 \end{bmatrix}u; \quad \boldsymbol{x}(0) = \boldsymbol{0}.$$

There is a requirement $r(t)=1$ on the element x_1. Find $u(t)$ and $x_1(t)$ if the integral $I = \int_0^T \{u^2+(r-x_1)^2\}\,dt$ is to be minimised, T being fixed. Obtain as a particular case the solution to the problem when

$T \to \infty$. (Suggestions: *Either* consider r as a state-vector element, $x_3 = r$, so that $\dot{x}_3 = 0$, *or* change the state-vector to $\boldsymbol{y}(t)$ where $y_1 = 1 - x_1$, $y_2 = x_2$. Better still use each of these methods in turn to decide which is the best!)

3. A simple plant has a transfer function relating output $V(s)$ to input $U(s)$ by

$$V(s) = \frac{8}{(s+1)(s+3)} U(s)$$

The system is inert at $t = 0$ and the requirement $r(t)$ on $v(t)$ is unit step function.

(i) Find the value $u(\infty)$ of $u(t)$ required to make $v(t) = r(t)$ under steady-state conditions.

(ii) Find $u(t)$ if $I = \int_0^\infty [\{r(t) - v(t)\}^2 + \frac{2}{5}\{u(t) - u(\infty)\}^2]\, dt$ is to be minimised and calculate its minimum value.

4. Return to Example 2 considering only the case $T \to \infty$. Obtain the solution for the co-state-vector $\boldsymbol{k}(t)$ as well as for $\boldsymbol{y}(t)$. Show that there is a unique, time-independent, symmetric matrix $\boldsymbol{N}$ such that $\boldsymbol{k}(t) = \boldsymbol{N}.\boldsymbol{y}(t)$. Show further that this matrix satisfies the Riccati equation, eqn. 11.26.

Substituting the values found in Example 2 for $\boldsymbol{y}(t)$ and $\boldsymbol{u}(t)$, evaluate the minimum value of I by direct integration and show that this value is also the value of $\frac{1}{2}\boldsymbol{y}'(0).\boldsymbol{N}.\boldsymbol{y}(0)$.

Chapter 12

Optimisation in the presence of amplitude constraints

12.1 Introduction

This Chapter is chiefly concerned with what may be called 'Pontryagin theory'. Pontryagin's Maximum Principle was first published in 1956 in a joint paper with Boltianskii and Gamkrelidze. The basic principles of this paper arise from what is, to the average engineering student, a rather sophisticated approach to dynamic systems in general which, for reasons of space, we shall not present. Only working rules will be given for the use of Pontryagin theory in connection with optimisation problems in which amplitude constraints apply to the input vector $\boldsymbol{u}(t)$, with particular attention to minimum-time and minimum-fuel problems.

Although the Pontryagin approach is basically the minimisation of a terminal-state function, whereas the calculus of variations approach of the last Chapter is essentially concerned with minimising an integral, it has already been noted in Section 10.3 that these two types of problem are readily interchangeable: in order to maintain continuity of presentation, the integral-minimisation approach will be maintained in the earlier part of the Chapter and a subsequent comparison will be made with the Pontryagin approach.

12.2 Amplitude constraints on the input vector

Such constraints, taking the general form

$$\boldsymbol{u}_{max} \geqslant \boldsymbol{u}(t) \geqslant \boldsymbol{u}_{min}, \text{ implying } u_{max,r} \geqslant u_r(t) \geqslant u_{min,r} \quad (r = 1, 2, \ldots m) \tag{12.1}$$

during the control interval $t_0 \leqslant t \leqslant t_1$, may be enforced in order to avoid overloading various parts of the plant or may be necessitated by the physical limitations of the inputs themselves: thus a jet engine has a certain maximum thrust, a 12V battery cannot give more than 12V etc. Whatever the reason for these constraints, their existence often means that the input functions determined by the solution of some optimisation problem by the methods of the calculus of variations exceed the range of these constraints and are therefore impracticable. One possible method of dealing with this difficulty is through the use of so-called *penalty functions*, added to the integrand of the integral to be minimised. For instance, selecting the rth constraint in eqn. 12.1, it will be appreciated that the function

$$|2u_r(t) - u_{max,r} - u_{min,r}|/(u_{max,r} - u_{min,r})$$

lies between 0 and 1 if $u_r(t)$ lies within the permitted range but is greater than unity if $u(t)$ is outside the permitted range. It follows that if this function is raised to a high positive power, the result will be a function which tends to zero when $u_r(t)$ is in the permitted range but which will rise very sharply to high values as soon as $u_r(t)$ transgresses either of its limits. The inclusion of such a function–a penalty function –in the integrand of the integral to be minimised will clearly lead to a solution for which $u_r(t)$ tends to obey its constraint equation. However, it is obvious that the addition of possibly m such functions (one for each element of $\boldsymbol{u}$ under constraint,) will not assist mathematical tractability! For this reason penalty functions are not widely used in constraint problems.

Before proceeding, it will be convenient to summarise the relevant results of the last Chapter:

If it is required to minimise $I = \int_{t_0}^{t_1} G(\boldsymbol{x}, \boldsymbol{u}, t)dt$ in the presence of the equality constraint $\dot{\boldsymbol{x}} = \boldsymbol{f}(\boldsymbol{x}, \boldsymbol{u}, t)$, the augmented integral is formed as

$$I_a = \int_{t_0}^{t_1} [G(\boldsymbol{x}, \boldsymbol{u}, t) + \boldsymbol{k}'(t)\{\boldsymbol{f}(\boldsymbol{x}, \boldsymbol{u}, t) - \dot{\boldsymbol{x}}\}]\, dt \equiv \int_{t_0}^{t_1} F(\dot{\boldsymbol{x}}, \boldsymbol{x}, \boldsymbol{u}, \boldsymbol{k}, t)dt$$

and conditions necessary for the minimisation are obtained as:

$$\frac{\partial F}{\partial \boldsymbol{u}} = 0, \text{ the } \textit{control} \text{ equation;} \quad \frac{\partial F}{\partial \boldsymbol{k}} = 0, \text{ the } \textit{state} \text{ equation } \dot{\boldsymbol{x}} = \boldsymbol{f};$$

$$\frac{\partial F}{\partial \boldsymbol{x}} = \frac{d}{dt}\frac{\partial F}{\partial \dot{\boldsymbol{x}}}, \text{ the } \textit{co-state} \text{ equation.}$$

Moreover, if t_0 and t_1 are fixed, then, both at $t = t_0$ and at $t = t_1$, either x_r $(r = 1, 2, \ldots n)$ is given, or x_r is free, in which case $k_r = 0$.

12.2.1 The function H

We now define a new scalar function H as

$$H(\boldsymbol{x}, \boldsymbol{u}, \boldsymbol{k}, t) \triangleq G(\boldsymbol{x}, \boldsymbol{u}, t) + \boldsymbol{k}'(t)\boldsymbol{f}(\boldsymbol{x}, \boldsymbol{u}, t) \tag{12.2}$$

$$= F(\dot{\boldsymbol{x}}, \boldsymbol{x}, \boldsymbol{u}, t) + \boldsymbol{k}'(t)\dot{\boldsymbol{x}} \tag{12.3}$$

since $F = G + \boldsymbol{k}'(\boldsymbol{f} - \dot{\boldsymbol{x}})$.

Using eqn. 12.3 in the form $F = H - \boldsymbol{k}'\dot{\boldsymbol{x}}$, we may express the above necessary conditions in terms of H instead of F: we deduce

$$\frac{\partial H}{\partial \boldsymbol{u}} = 0, \text{ the control equation; } \frac{\partial H}{\partial \boldsymbol{k}} = \dot{\boldsymbol{x}}', \text{ the state equation}$$

$$\frac{\partial H}{\partial \boldsymbol{x}} = -\dot{\boldsymbol{k}}', \text{ the co-state equation.} \tag{12.4}$$

We may note in passing an interesting property of H. Since it is a function of $\boldsymbol{x}, \boldsymbol{u}, \boldsymbol{k}$ and t, its total time-derivative is

$$\frac{dH}{dt} = \frac{\partial H}{\partial \boldsymbol{x}}\dot{\boldsymbol{x}} + \frac{\partial H}{\partial \boldsymbol{u}}\dot{\boldsymbol{u}} + \frac{\partial H}{\partial \boldsymbol{k}}\dot{\boldsymbol{k}} + \frac{\partial H}{\partial t},$$

or, using eqn. 12.3,

$$= -\dot{\boldsymbol{k}}'\dot{\boldsymbol{x}} + 0 + \dot{\boldsymbol{x}}'\dot{\boldsymbol{k}} + \frac{\partial H}{\partial t}, \text{ where } \dot{\boldsymbol{k}}'\dot{\boldsymbol{x}} = \text{scalar} = \dot{\boldsymbol{x}}'\dot{\boldsymbol{k}},$$

$$= \frac{\partial H}{\partial t}$$

Hence if H is not *explicitly* a function of t, so that its partial time-derivative vanishes, then also its total time-derivative vanishes, i.e. *H is then constant throughout the control interval provided an optimal trajectory is being followed.*

The key property of the function H, however, whether or not it is explicitly time-dependent, may be stated as follows:

> *If an optimal trajectory is followed, then, at every instant of the control interval, the value of H is to be minimised.*

This rule is to be followed whether or not inequality constraints such as those of eqn. 12.1 are applied. If such constraints are in force, then it may be that the optimising conditions for eqn. 12.4, which are turning-value conditions, cannot be satisfied within the prescribed ranges of the constrained variables; even if they can, they may not correspond to the least value of H, which may occur at one or other of the extremes of the prescribed ranges. The minimisation of H at every instant, which is required by the rule above, is of course to be achieved by selecting, at every instant, suitable values of the elements of the input-vector $\boldsymbol{u}(t)$. As an illustration of the rule we shall first consider the problem of control-time minimisation, with $\boldsymbol{x}(t_0)$ and $\boldsymbol{x}(t_1)$ given.

12.2.2 Control-time minimisation

Since the control-time is

$$t_1 - t_0 = \int_{t_0}^{t_1} dt$$

minimisation of the control-time requires $G(\boldsymbol{x}, \boldsymbol{u}, t) \equiv 1$. Hence, from eqn. 12.2

$$H = 1 + \boldsymbol{k}'\boldsymbol{f}(\boldsymbol{x}, \boldsymbol{u}, t)$$

Consider as a particular case the linear, autonomous system for which $\dot{\boldsymbol{x}} = \boldsymbol{A}\boldsymbol{x} + \boldsymbol{B}\boldsymbol{u}$, so that $H = 1 + \boldsymbol{k}'\boldsymbol{A}\boldsymbol{x} + \boldsymbol{k}'\boldsymbol{B}\boldsymbol{u}$. If $\boldsymbol{u}$ is not constrained in the amplitude of its elements, it is clear that since H is linear in the elements of $\boldsymbol{u}$, the minimisation of H (to $-\infty$) will be achieved by making the elements of $\boldsymbol{u}$ equal to $\pm\infty$ depending on whether their coefficients in H are instantaneously negative or positive. This merely indicates that we can minimise the control interval (to zero) provided we are allowed to use sufficiently large inputs. If, however, we introduce the constraints of eqn. 12.1, the control equation, $\dfrac{\partial H}{\partial \boldsymbol{u}} = 0$, clearly has no solution, (apart from the rather trivial solution $\boldsymbol{k} = \boldsymbol{0}$, i.e. $H = 1$, and hence $\dot{\boldsymbol{x}}' = \dfrac{\partial H}{\partial \boldsymbol{k}} = \boldsymbol{0}'$, which makes it very difficult to get from $\boldsymbol{x}_0$ to $\boldsymbol{x}_1$ except in infinite time!). If, however, we substitute the 'H must be minimised' rule for the control equation, then, at every instant, any element of $\boldsymbol{u}(t)$, say $u_r(t)$, must be made equal to $u_{min,r}$ or $u_{max,r}$ depending on whether, at that instant, the coefficient of $u_r(t)$ in H, namely the rth element of $\boldsymbol{k}'(t)\boldsymbol{B}$, is positive or negative. Hence the 'control strategy' must necessarily be of what is termed the 'bang-bang' type, each element $u_r(t)$ of the input-vector being instantaneously switched from its minimum permissible value $u_{min,r}$ to its maximum permissible value $u_{max,r}$ (or *vice versa*) whenever, due to the time-variation of $\boldsymbol{k}(t)$, $\boldsymbol{k}'(t)\boldsymbol{b}_r$ goes through zero from positive to negative (or vice versa).

12.2.2.1 Example

$$A = \begin{bmatrix} 0 & 1 \\ 0 & 0 \end{bmatrix}, B = \begin{bmatrix} 0 \\ 1 \end{bmatrix}; x_0 = \begin{bmatrix} 0 \\ 0 \end{bmatrix}, x_1 = \begin{bmatrix} 1 \\ 0 \end{bmatrix}; -1 \leqslant u \leqslant 2$$

In physical terms, since $\dot{x}_2 = \ddot{x}_1 = u$, a particle of unit mass is capable of unresisted motion in one dimension (x_1) under the influence of a force u bounded by the limits (-1) and ($+2$). The particle is to be brought from rest at $x_1 = 0$ to rest at $x_1 = 1$ in minimum time. Find $u(t)$.

Solution

We have $H = 1 + k'Ax + k'Bu = 1 + k_1x_2 + k_2u$

To minimise H with respect to u therefore requires $u = -1$ if k_2 is positive, $u = 2$ if k_2 is negative, the switching instants being when $k_2 = 0$. Now $k(t)$ is controlled by the co-state equation, $\dfrac{\partial H}{\partial x} = -\dot{k}'$ or, in this case $\dot{k} = -A'k$, i.e. $\dot{k}_1 = 0, \dot{k}_2 = -k_1$. We therefore postulate $k_1 = a, k_2 = b - at$, with a, b constants. There can therefore be only one switching instant, $t = b/a$, though of course this value of time may fall outside the control interval.

However, if u is constant in any subinterval, the particle either accelerates or decelerates as $u \gtrless 0$. Since there is to be only one switching instant and the particle starts and finishes at rest, this switching instant must physically exist within the control interval. Since, moreover, the distance travelled is to be positive, ($+1$), the period of acceleration must precede the period of deceleration, otherwise the velocity would be negative within the control interval. Since the acceleration is 2 and the deceleration is 1 and there is no net change in velocity, the acceleration period, say T_1, must be one half of the deceleration period, say T_2. The distance travelled while accelerating from rest is $\frac{1}{2}.2.T_1^2$ and while decelerating to rest is $\frac{1}{2}.1.T_2^2$. Hence $T_1^2 + T_2^2/2 = 1$ and $T_2 = 2T_1$ so that $T_1 = 1/\sqrt{3}$, $T_2 = 2/\sqrt{3}$ and $T_1 + T_2 =$ minimum control-time $= \sqrt{3}$.

The solution of this simple 2-dimensional problem may of course be illustrated either by velocity-time and displacement-time graphs or by a velocity-displacement graph which, since $x_2 = \dot{x}_1$, is also in this case a state diagram. (Fig. 12.1)

With constant $u, x_2 = ut + c_1$ and hence $x_1 = ut^2/2 + c_1t + c_2$, where c_1, c_2 are integration constants. Eliminating t produces $x_1 = x_2^2/2u + c_2 - c_1^2/2u = x_2^2/2u + k$, so that the state trajectories are parabolas; putting $u = 2, u = -1$, gives two families of parabolas as shown, members of one family only differing by their values of k, and these imply a horizontal shift. The direction of increasing time is

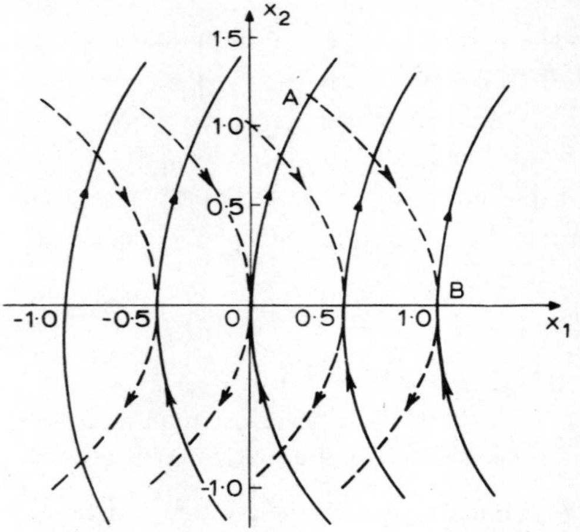

Fig. 12.1 Trajectories for example of 12.2.2.1. Full-line parabolae correspond to $u = 2$, broken-line to $u = -1$. Arrows indicate the direction of increasing time

obtained from $\dot{x}_2 = u$, so that, with u positive (negative), x_2 increases (decreases) with increasing time. It is obvious that the only path from O to B with only one switch is by switching at A. Since the parabola OA is $x_1 = x_2^2/4$ and the parabola AB is $x_1 - 1 = -x_2^2/2$, their intersection at A is the point $(1/3, 2/\sqrt{3})$. Since on any parabola, $x_2 = ut + c_1$, the time taken to travel along an arc is the increase in x_2 divided by u: hence the time taken from O to A is $1/\sqrt{3}$ and from A to B is $2/\sqrt{3}$, confirming earlier results.

12.2.2.2 Comments

(i) The example is admittedly trivial, deliberately so, in order that the results obtained by the H-minimisation technique may be easily compared with those obtained from a commonsense approach.

(ii) It is important to note that the H-minimisation technique, (which, as will be seen later, is identical with the Pontryagin Maximum Principle,) only tells us two things: (*a*) the control must be of the bang-bang type, (*b*) there cannot be, in this problem, more than one switching instant; but it does not solve the problem. For, without further logical investigation, we do not know at which extreme value of u to begin, nor, in this problem, do we know the switching instant; this can only be found by adjusting its value until the trajectory goes through its desired end-points. It must be appreciated that in a more complex problem with larger values of m and n this difficulty may be greatly increased: although in a linear autonomous problem, if the matrix A has distinct eigenvalues, then it may be shown that no element of the vector $u(t)$ need be switched more than $(n-1)$ times, this still leaves the possibility of $m(n-1)$ switching instants, the location of which will demand considerable trial and error investigation.

(iii) Analysing the optimal solution found for this problem, we know that $H = 1 + k_1 x_2 + k_2 u$ and that in the first subperiod $(0 \leqslant t \leqslant 1/\sqrt{3})$, $u = 2$, while in the second subperiod $(1/\sqrt{3} \leqslant t \leqslant \sqrt{3})$, $u = -1$. Also $k_1 = a$, $k_2 = b - at$, and since $k_2 = 0$ determines the switching instant, it follows that $k_2 = a(1/\sqrt{3} - t)$. Also, x_2 vanishes at $t = 0$ and at $t = \sqrt{3}$ and $\dot{x}_2 = u$: hence $x_2 = 2t$ during the first subperiod and $x_2 = (\sqrt{3} - t)$ during the second subperiod. Making these substitutions we find that during the first subperiod $H = 1 + 2at + 2a(1/\sqrt{3} - t) = 1 + 2a/\sqrt{3}$, while during the second subperiod $H = 1 + a(\sqrt{3} - t) - - a(1/\sqrt{3} - t) = 1 + 2a/\sqrt{3}$ also, thus verifying the property, noted in Section 12.2.1, that when H is not explicitly a function of time, it remains constant during the whole of the control interval.

12.2.3 Fuel minimisation

The rate of fuel consumption is supposed at first, generally, to be $\phi = \phi(\boldsymbol{x}, \boldsymbol{u}, t)$. The integral $\int_{t_0}^{t_1} \phi . dt$ has to be minimised so that for this problem $G = \phi$, the minimisation being subject to the equality constraint $\dot{x} = f(x, u, t)$ and the inequality constraints $\boldsymbol{u}_{max} \geqslant \boldsymbol{u}(t) \geqslant \boldsymbol{u}_{min}$. A frequent simplification occurs where ϕ is independent not merely of t but also of $\boldsymbol{x}$, so that $\phi = \phi(\boldsymbol{u})$. A further simplification of form occurs if it is assumed that any fuel-consuming element of $\boldsymbol{u}$, say u_r, is associated with a rate of fuel-consumption proportional to $|u_r|$ (implying that the fuel consumption is unchanged by reversing u_r) so that ϕ takes the form $\Sigma\, \alpha_r |u_r|$. Whether or not any of these simplifications apply, certain conditions for minimisation are obtainable from the same prinicple as before: H must be minimised at every instant of the control period. In this case $H = \phi + \boldsymbol{k}'\boldsymbol{f}$.

The method is illustrated with the same very simple plant equation as before and the same limits on u as in the last example, but with elastic limits on the terminal values of $\boldsymbol{x}$.

12.2.3.1 Example

The plant is represented by $\dot{x}_1 = x_2, \dot{x}_2 = u$, i.e. $A = \begin{bmatrix} 0 & 1 \\ 0 & 0 \end{bmatrix}$,

$B = \begin{bmatrix} 0 \\ 1 \end{bmatrix}$ and $2 \geqslant u \geqslant -1$. The rate of fuel consumption is proportional to $|u|$ so that fuel-minimisation is equivalent to minimising $\int |u| dt$ over some interval $0 \leqslant t \leqslant T$ where T is unspecified. The initial state is $x_0 = \begin{bmatrix} a_0 \\ b_0 \end{bmatrix}$, the final state is $x_1 = \begin{bmatrix} a_1 \\ b_1 \end{bmatrix}$.

Solution:
We have $H = |u| + k'(Ax + Bu) = |u| + k_1x_2 + k_2u$. Considering this as a function of u to be minimised, we note that the slope (or derivative) of the first term is ± 1 as $u \gtrless 0$, that the second term is constant, and that the slope of the third term is k_2. The graph of H against u therefore consists of two straight lines, one of slope $(-1 + k_2)$ covering the range $u < 0$, the other of slope $(1 + k_2)$ covering the range $u > 0$, the lines intersecting at $u = 0$, $H = k_1x_2$. If both slopes are negative, i.e. if $k_2 < -1$, H is clearly minimised by making u equal its upper limit of $+2$. Similarly, if both slopes are positive, i.e. if $k_2 > +1$, H is minimised by making u equal its lower limit of -1. But if $-1 < k_2 < 1$, H is clearly minimised by making $u = 0$, since the slope to the left of this value will be negative, while the slope to the right will be positive. (See Fig. 12.1)

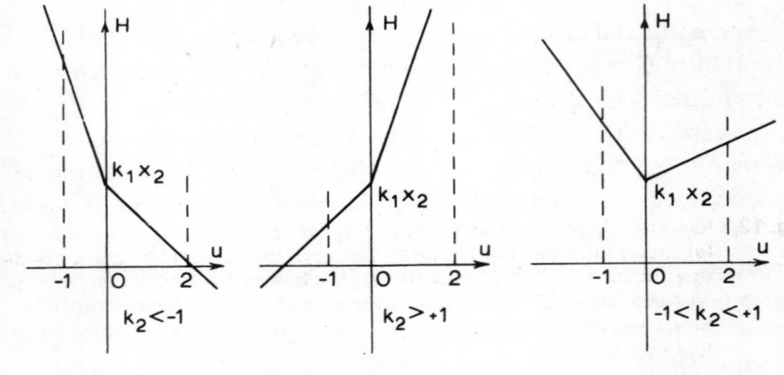

Fig. 12.2 Minimising $H = |u| + k_1x_2 + k_2u$

Now $k(t)$ is controlled by the same equation, $\dfrac{\partial H}{\partial x} = -\dot{k}'$, as in the previous example, resulting in $k_1 = a$, $k_2 = b - at$. Since k_2 varies linearly with time, it must, at two distinct times, cross the critical values $k_2 = \pm 1$, but these times may, or may not, lie within the control interval: these times are unknown and so is the order in which the

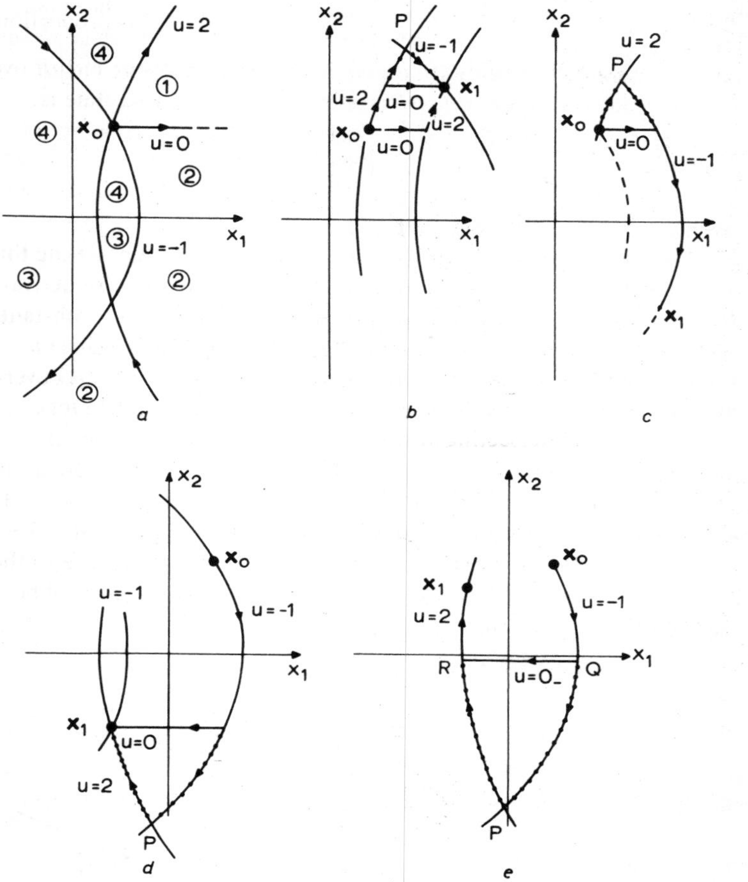

Fig. 12.3 Minimum-fuel and minimum-time trajectories
(*a*) shows the zoning of the phase-plane; in (*b*), (*c*), (*d*), (*e*) corresponding to $\mathbf{x}_1$ in zone 1, 2, 3, 4, respectively, thick lines, continuous or broken, represent minimum-fuel trajectories; dotted lines represent minimum-time trajectory segments replacing the horizontal parts of the minimum-fuel trajectories

critical values are crossed since this depends upon the sign of a. We can say, however, that the behaviour of $u(t)$ must follow either the sequence $(2, 0, -1)$ or $(-1, 0, 2)$, which we may call a 'bang-O-bang' type of control.

We shall again illustrate the specific solution by a state-space diagram, Fig. 12.3, remembering that in this case the transfer from a $u = 2$ parabola to a $u = -1$ parabola, or *vice versa*, takes place through an interim stage corresponding to $u = 0$, i.e. $\dot{x}_2 = 0$, namely a horizontal trajectory in the diagram. To avoid too many diagrams we shall assume that b_0, the initial value of x_2, is positive. Fig. 12.3*a* shows such a starting point with the three trajectories through it corresponding to $u = -1, u = 0, u = 2$. These three trajectories delimit certain numbered zones in the state-plane, for it will be found that the nature of the optimal trajectory depends upon the zone-location of the terminal point $\underline{x}_1$. It must be remembered that:

(*a*) the most general sequence for u is either $(2, 0, -1)$ or $(-1, 0, 2)$ but, since either or even both switching instants may fall outside the control interval, 2-term sequences $(2, 0)$, $(0, -1)$, $(0, 2)$, $(-1, 0)$ or even 1-term, (2), (0), (-1), may provide a solution:

(*b*) since the rate of fuel-consumption is $|u|$ and $u = \dot{x}_2$ it follows that the fuel consumed on any part of a constant-u trajectory equals the increase in x_2 if $u > 0$, the decrease in x_2 if $u < 0$ and is of course zero if $u = 0$;

(*c*) although we are not primarily interested in the time taken, as in the previous example, the time taken on a parabolic arc is the increase in x_2 divided by 2 or (-1) as the case may be; if $u = 0, \dot{x}_1 = x_2 =$ constant and the time taken is therefore the increase of x_1 divided by the value of x_2. (Note that if $x_2 > 0, x_1$ increases with time, but if $x_2 < 0, x_1$ decreases as time increases.)

Case 1: $\boldsymbol{x}_1$ in Zone 1, Fig. 12.3*b*. Two 2-term solutions are shown, $u = (2, 0)$, (full-line), and $u = (0, 2)$, (broken-line). In fact an infinite number of solutions is possible, *with the same fuel consumption*, in which u alternates between $u = 0$ and $u = 2$ with as many steps as we please, provided only that the resultant trajectory is confined to the space bounded by the two 2-term trajectories shown. The fuel consumed is $(b_1 - b_0)$ for all these solutions, but the time taken is least for the u-sequence $(2, 0)$, which is therefore probably preferable. The minimum-*time* trajectory, through P, is associated with a larger fuel consumption $(2x_{2P} - b_1 - b_0)$, which exceeds $(b_1 - b_0)$ since $x_{2P} > b_1$. Any usage of a $u = -1$ parabola, by (*b*) above, will increase the fuel consumption.

Case 2: $\boldsymbol{x}_1$ in Zone 2, Fig. 12.3*c*. The only 2-term u-sequence permissible is $(0, -1)$. The reverse sequence $(-1, 0)$ is impossible since it would involve x_1 increasing with time when x_2 is negative. The use of a $u = 2$ parabola necessarily increases the fuel consumption. An infinite number of solutions with the same minimum fuel consumption again exist, consisting of alternations between $u = 0$ and $u = -1$, provided that the resulting trajectory lies between $x_2 = b_0$ and $x_2 = b_1$ and also to the right of the $u = -1$ parabola through $\boldsymbol{x}_0$ in the upper half-plane and to the right of the $u = -1$ parabola through $\boldsymbol{x}_1$ in the lower half-plane. The minimum fuel consumption is $(b_0 - b_1)$ and *for this consumption* the least-time trajectory is given by the 2-term sequence $u = (0, -1)$. The minimum-*time* trajectory through P clearly uses more fuel.

Case 3: $\boldsymbol{x}_1$ in Zone 3, Fig. 12.3*d*. Only the sequence $(-1, 0)$ gives minimum fuel consumption, though again repeated alternations of u between these values are permissible, again with certain constraints on the trajectory. The fuel consumption on the trajectory shown, $u = (-1, 0)$, is again $(b_0 - b_1)$ and is again added to by following the minimum-*time* trajectory through P.

Case 4: $\boldsymbol{x}_1$ in Zone 4, Fig. 12.3*e*. Here we must use the 3-term sequence $u = (-1, 0, 2)$, though again an infinite number of other trajectories give the same consumption. In order to minimise the consumption the negative value of x_2 from Q to R should be as small as possible in magnitude, but as it tends to zero, the time from Q to R, by (*c*) above, tends to infinity. Clearly, in practice, a compromise between minimum-time and minimum-fuel must be decided upon and the path QR will lie between P and the axis of x_1. The fuel consumption will then be $(b_0 + b_1 - 2x_{2QR})$, with $0 > x_{2QR} > x_{2P}$.

12.2.3.2 Comments

(i) Although in this simple example an infinite number of trajectories give the same minimum fuel consumption, this in no way contradicts the results of our theory, namely that minimum fuel consumption may be achieved (not necessarily exclusively) by using one or other of the u-sequences given in (*a*) above.

(ii) We have dealt with this simple problem fairly thoroughly in order to drive home the point made earlier, that the *H*-minimising technique only gives a limited amount of information, in this case that the control strategy is at most of the bang-O-bang type. The rest of the solutions obtained above are in substance trial and error procedures conditioned by the terminal values of the problem. In this case, armed with an analytical solution for the trajectory equations and a diagrammatic interpretation, the trial and error processes were avoidable, but this situation is exceptional. Nevertheless the information given by the *H*-minimisation technique must not be underestimated either: think how much more trial and error would be necessary in this problem if we did not know at the start that the control strategy was to be of bang-O-bang type!

12.3 The Maximum Principle of Pontryagin

It is supposed that the plant considered is representable by the state equation $\dot{\boldsymbol{x}} = \boldsymbol{f}(\boldsymbol{x}, \boldsymbol{u}, t)$; starting from a state $\boldsymbol{x} = \boldsymbol{x}_0$ at $t = t_0$ it is required to find the control strategy [i.e. the value of $\boldsymbol{u}(t)$] so that some linear function of the end-state, $\boldsymbol{x} = \boldsymbol{x}_1(t)$ at $t = t_1$, shall be minimised. This function is called the Pontryagin function and is here denoted by

$$P = \boldsymbol{b}'.\boldsymbol{x}_1 = \sum_{r=1}^{n} b_r x_{1r}$$

where $\boldsymbol{b}$ is some given column of coefficients. For the present, $\boldsymbol{x}_1$ will be considered free, i.e. none of its components are specified.

Pontryagin's Maximum Principle states that if the control strategy is optimal, in the sense that it minimises P, then it also maximises a scalar *Hamiltonian function* $H_0(\boldsymbol{x}, \boldsymbol{u}, \boldsymbol{p}, t)$ with respect to $\boldsymbol{u}(t)$ at every instant of the control interval. The vector $\boldsymbol{p} = \boldsymbol{p}(t)$ which appears as one of the arguments of H_0 is defined by

$$\dot{\boldsymbol{p}} = -\left(\frac{\partial \boldsymbol{f}}{\partial \boldsymbol{x}}\right)' \boldsymbol{p} \quad \text{or} \quad \dot{p}_r = -\sum_{s=1}^{n} p_s \frac{\partial f_s}{\partial x_r}, \qquad (r = 1, 2, \ldots n) \quad (12.5)$$

with the terminal condition

$$\boldsymbol{p}(t_1) \equiv \boldsymbol{p}_1 = -\boldsymbol{b} \quad \text{or} \quad p_{1r} = -b_r \qquad (r = 1, 2, \ldots, n) \quad (12.6)$$

and $\boldsymbol{p}$ is often called the *momentum vector* since its elements are momenta in the mechanical interpretation of the theory.

With $\boldsymbol{p}$ thus completely defined (since $\boldsymbol{f}$ and $\boldsymbol{b}$ are both known), the Hamiltonian function H_0, which has the dimensions of energy, is defined by

$$H_0 = \boldsymbol{p}'.\boldsymbol{f} = \boldsymbol{f}'.\boldsymbol{p} = \sum_{r=1}^{n} p_r f_r \qquad (12.7)$$

Differentiating this equation partially with respect to $\boldsymbol{p}$ gives

$$\frac{\partial H_0}{\partial \boldsymbol{p}} = \boldsymbol{f}' = \dot{\boldsymbol{x}}', \text{ using the state equation,} \qquad (12.8)$$

and differentiating it partially with respect to $\boldsymbol{x}$ gives

$$\frac{\partial H_0}{\partial \boldsymbol{x}} = \boldsymbol{p}'\frac{\partial \boldsymbol{f}}{\partial \boldsymbol{x}} = -\dot{\boldsymbol{p}}', \text{ using eqn. 12.5} \tag{12.9}$$

These last two relations, eqns. 12.8 and 12.9, are known as the *Hamiltonian canonical equations.*

The solution of the problem is subject to the 2-point boundary conditions

$$\boldsymbol{x} = \boldsymbol{x}_0 \text{ at } t = t_0 \quad \text{and} \quad \boldsymbol{p} = -\boldsymbol{b} \text{ at } t = t_1 \tag{12.10}$$

The process of solution may be summarised as follows: maximise H_0 at every instant by choosing a suitable $\boldsymbol{u}(t)$ and thus obtain, if possible, an optimal vector $\boldsymbol{u}(\boldsymbol{p}, t)_{opt}$ which, due to the form of H_0, will be a function of $\boldsymbol{p}$. If this value of $\boldsymbol{u}(t)$ is substituted in the canonical equations (eqns. 12.8 and 12.9) where it appears implicitly through $\boldsymbol{f}(\boldsymbol{x}, \boldsymbol{u}, t)$, these $2n$ first-order equations may be solved for $\boldsymbol{x}(t)_{opt}$ and $\boldsymbol{p}(t)_{opt}$ subject to the $2n$ boundary conditions of eqn. 12.10.

12.3.1 Non-free end state

If the final state $\boldsymbol{x}$ is *partially* constrained (if it is totally constrained, i.e. if $\boldsymbol{x}_1$ is specified, P is a constant and the problem of its minimisation does not arise,) and if these constraints are, quite generally, expressed by, say,

$$\boldsymbol{c}(\boldsymbol{x}_1) = \boldsymbol{0} \quad \text{or} \quad c_r(\boldsymbol{x}_1) = 0, r = 1, 2, \ldots, N < n$$

then the Pontryagin function takes the form

$$P = \boldsymbol{b}'\boldsymbol{x}_1 + \boldsymbol{\mu}'\boldsymbol{c} \tag{12.11}$$

where $\boldsymbol{\mu}$ is a Lagrange multiplier vector of order N. The canonical equations are still valid but are now subject to the boundary conditions

$$\boldsymbol{x} = \boldsymbol{x}_0 \text{ at } t = t_0 \qquad n \text{ equations} \tag{12.12a}$$

$$\boldsymbol{p} = -\boldsymbol{b} - (d\boldsymbol{c}/d\boldsymbol{x}_1)'\boldsymbol{\mu} \text{ at } t = t_1 \qquad n \text{ equations} \tag{12.12b}$$

$$\boldsymbol{c}(\boldsymbol{x}) = \boldsymbol{0} \text{ at } t = t_1 \qquad N \text{ equations} \tag{12.12c}$$

giving a total of $(2n + N)$ boundary conditions to determine the N elements of $\boldsymbol{\mu}$ and to solve the $2n$ canonical equations.

If, as often happens, the constraints on the end-state take the form $c_r(\boldsymbol{x}_1) = x_{1r} - X_r = 0$, i.e. if $x_r = X_r$ is given at $t = t_1$, then any such given elements of the final state may be associated, in the original form of $P = \boldsymbol{b}'\boldsymbol{x}_1$, with an arbitrary b_r, since $b_r x_{1r}$ will then have a known constant value, and will therefore be irrelevant to the problem of maximising P. Since also, with this form of the c_r, dc_s/dx_r will be zero unless

$s = r$ when its value is unity, it follows that, in eqn. 12.12*b*, if x_{1r} is given, then $p_{1r} = -b_r - \mu_r$ but if x_{1r} is free then $p_{1r} = -b_r$. The boundary conditions of eqn. 12.12 may then be stated more simply, by dispensing with the Lagrange multiplier, as

$$x = x_0 \quad \text{at } t = t_0 \qquad (n \text{ equations}) \qquad (12.13a)$$

$$p_r = -b_r \quad \text{at } t = t_1 \text{ provided } x_{1r} \text{ is free } [(n-N) \text{ equations}] \qquad (12.13b)$$

$$x_r = X_r \quad \text{at } t = t_1 \text{ where } X_r \text{ is given } (N \text{ equations}) \qquad (12.13c)$$

giving $2n$ boundary conditions for the solution of the canonical equations. In this case, moreover, P may be taken as $\Sigma\, b_r x_{1r}$, the summation being limited to the *free* elements of x_1.

12.3.2 Non-linear Pontryagin function

In spite of the form $P = b'x_1$, the Pontryagin theory is not limited to minimising linear functions of the end state. Indeed if we wish to minimise $K(x_1)$ where K is differentiable, let us create an additional state element, $x_{n+1} = K(x)$ with dynamic equation $\dot{x}_{n+1} = f_{n+1} = \dfrac{dK}{dx}\dot{x} = \dfrac{dK}{dx} f$, in which x denotes the *un*augmented state vector. We then write $P = x_{n+1}(t_1)$ so that $b_{n+1} = 1$ and $b_1 = b_2 = \ldots = b_n = 0$.

12.4 Reconciling the two approaches

We revert to the problem presented in Sections 12.2–12.2.1: to find $u(t)$ so as to minimise $I = \int_{t_0}^{t_1} G(x, u, t)dt$ subject to the state equation $\dot{x} = f(x, u, t)$. The augmented integral I_a was formed with integrand $F = G + k'(f - \dot{x})$ and necessary conditions were deduced for the minimisation of I, the control equation, the state equation and the co-state equation. On expressing these equations in terms of $H = G + k'f = F + k'\dot{x}$ and on substituting the H-minimisation principle for the control equation, we obtained as necessary conditions for the solution:

H is to be minimised at every instant by choice of a suitable $u(t)$;

$\dfrac{\partial H}{\partial k} = \dot{x}'$, (the state equation) and $\dfrac{\partial H}{\partial x} = -\dot{k}'$, (the co-state equation).

As far as the boundary conditions were concerned, both at t_0 and at t_1 either x_r is given, or x_r is free and then $k_r = 0$.

Consider the same problem solved by the Pontryagin method. First we introduce an additional state element $x_{n+1}(t) = \int_{t_0}^{t} G(x, u, \tau)d\tau$, so that $\dot{x}_{n+1} = G(\boldsymbol{x}, \boldsymbol{u}, t) = f_{n+1}(\boldsymbol{x}, \boldsymbol{u}, t)$. (Note that here, as in the last Section, the symbols $\boldsymbol{x}$ and $\boldsymbol{f}$ refer to the original, *un*augmented vectors.)

Since the relevant Pontryagin function is simply $P = x_{n+1}(t_1)$, $b_{n+1} = 1$ and $b_1 = b_2 = \ldots = b_n = 0$. The Hamiltonian function H_0 is given by

$$H_0 \;=\; \boldsymbol{p}'\boldsymbol{f} + p_{n+1} f_{n+1}\,, \text{ where } \boldsymbol{p} \text{ excludes } p_{n+1}\,.$$

From eqn. 12.5, the dynamic equation for p_{n+1} is

$$\dot{p}_{n+1} \;=\; -\sum_{s=1}^{n+1} p_s \frac{\partial f_s}{\partial x_{n+1}} \;=\; 0$$

since *every* f_s, though a function of $\boldsymbol{x}$, is independent of x_{n+1}. Hence p_{n+1} is constant throughout the interval; but since $x_{n+1}(t_1)$ is free, it follows from eqn. 12.13*b* that $p_{n+1}(t_1) = -b_{n+1} = -1$. Hence $p_{n+1} = -1$ throughout and

$$H_0 \;=\; \boldsymbol{p}'\boldsymbol{f} - f_{n+1} \;=\; \boldsymbol{p}'\boldsymbol{f} - G(\boldsymbol{x}, \boldsymbol{u}, t)$$

[Compare $\quad H = \boldsymbol{k}'\boldsymbol{f} + G(\boldsymbol{x}, \boldsymbol{u}, t)$]

Moreover, having dealt with $\dot{p}_{n+1}$, eqn. 12.9 gives $\dot{\boldsymbol{p}}' = -\dfrac{\partial H_0}{\partial \boldsymbol{x}}$ (to be compared with $\dot{\boldsymbol{k}}' = -\dfrac{\partial H}{\partial \boldsymbol{x}}$). Similarly, since $H_0 = \boldsymbol{p}'\boldsymbol{f} - f_{n+1} = \boldsymbol{f}'\boldsymbol{p} - f_{n+1}$ it follows that $\boldsymbol{f}' = \dfrac{\partial H_0}{\partial \boldsymbol{p}} = \dot{\boldsymbol{x}}'$, to be compared with $\dfrac{\partial H}{\partial \boldsymbol{k}} = \dot{\boldsymbol{x}}'$.

It will be clear that, from these comparisons, we can convert from the H_0 equations to the H equations by the simple expedient of replacing H_0 by $-H$ and $\boldsymbol{p}$ by $(-\boldsymbol{k})$. In other words *the H function is the negative of the Hamiltonian function H_0 so that minimising H is identical with maximising H_0, while what we have defined as the co-state vector* $\boldsymbol{k}(t)$ *is the negative of the Pontryagin momentum vector* $\boldsymbol{p}(t)$.

Finally, as far as the boundary conditions are concerned, at $t = t_0$ the Pontryagin theory assumes $\boldsymbol{x} = \boldsymbol{x}_0$ (given). At $t = t_1$, the Pontryagin theory (by eqn. 12.13*b*) gives $p_r = -b_r = 0$ for any r in $1 \leqslant r \leqslant n$ for which x_r is free; this is to be compared with $k_r = 0$, if x_r is free, by the earlier method; since $\boldsymbol{k} = -\boldsymbol{p}$, these conditions are again identical. The two methods are therefore in complete agreement and only differ in the notation used.

We shall not here give a proof of the Pontryagin Maximum Principle (which, as just shown, justifies the earlier $\boldsymbol{H}$-minimisation technique); a proof may be found in a number of text-books or, of course, in the original paper.

12.5 Linear autonomous system: the possible and the impossible

In this section some aspects of time (or fuel) minimisation will be displayed which were not revealed in the simple examples so far given. For this purpose it will be useful first to normalise the input constraints of eqn. 12.1 and then to put the state-equation in a canonical form.

Write

$$u_{mean} \equiv \tfrac{1}{2}(u_{max} + u_{min}) \text{ and } \boldsymbol{\delta} \equiv \tfrac{1}{2}(u_{max} - u_{min})$$

so that

$$u_{max} = u_{mean} + \boldsymbol{\delta} \qquad \text{and } u_{min} = u_{mean} - \boldsymbol{\delta}.$$

Then, in eqn. 12.1

$$+\boldsymbol{\delta} \geqslant u(t) - u_{mean} \geqslant -\boldsymbol{\delta}$$

or, premultiplying by diag $(1/\delta_r)$, and with $\mathbf{1} \equiv$ a column of unity elements,

$$+\mathbf{1} \geqslant \text{diag}\,(1/\delta_r).\,\{u(t) - u_{mean}\} \geqslant -\mathbf{1}$$

Writing $w(t) \equiv \text{diag}\,(1/\delta_r).\,\{u(t) - u_{mean}\}$, i.e. $u(t) = u_{mean} +$ diag $\delta_r . w(t)$ we deduce

$$\mathbf{1} \geqslant w(t) \geqslant -\mathbf{1} \tag{12.14}$$

while substitution in the state equation $\dot{x} = Ax + Bu$ gives

$$\dot{x} = Ax + Bu_{mean} + B\,\text{diag}\,\delta_r .\,w$$

or

$$\dot{X} = AX + B\,\text{diag}\,\delta_r .\,w \text{ where } X = x + A^{-1}Bu_{mean} \tag{12.15}$$

on the supposition that A is nonsingular, i.e. has no zero eigenvalues. (Note that if $u_{mean} = 0$, i.e. if all the original constraints on the elements of u are symmetric about zero, this restriction is unnecessary and $x = X$.)

If we now suppose that the eigenvalues of A are distinct and that E is an associated eigenvector-assembly matrix, we may, putting $X = Ey$, obtain the canonical form of this state equation in the form

$$\dot{y} = \Lambda y + E^{-1}B\,\text{diag}\,\delta_r .\,w, \text{ where } \Lambda = \text{diag}\,\lambda_r \tag{12.16}$$

However, if A has any complex eigenvalues (occurring as conjugate complex pairs since A is real), the associated eigenvectors will also be complex and so therefore will be the new state-vector $y = E^{-1}X$. This clearly complicates the concept of trajectories in the (now complex) state-space! We may overcome this difficulty, at the cost of slightly modifying the diagonal nature of Λ, by the following artifice.

If A has one or more pairs of conjugate complex eigenvalues, reorder the eigenvalues if necessary in such a way that every such pair are a consecutive pair, for instance λ_c and its conjugate λ_{c+1}. The eigenvectors associated with this pair, say e_c and e_{c+1} may then be taken to be complex conjugate vectors. Now consider a matrix K formed from unity matrix I by changing the principal submatrix formed by its cth and $(c+1)$th rows *and* columns from $\begin{bmatrix} 1 & 0 \\ 0 & 1 \end{bmatrix}$ to $\begin{bmatrix} 1/2 & 1/2j \\ 1/2 & -1/2j \end{bmatrix}$, for every c such that λ_c, λ_{c+1} are a conjugate complex pair. Then on forming the matrix $F = EK$, the cth and $(c+1)$th columns of F will be $f_c = \frac{1}{2}(e_c + e_{c+1})$, which is real, and $f_{c+1} = \frac{1}{2j}(e_c - e_{c+1})$, which is also real. Hence F will be a real matrix, since, if any λ_s is real, $k_{ss} = 1$ and $f_s = e_s$ which is also real.

If we now make the substitution $X = F.z$ in eqn. 12.15, z will be real since F and X are real; also

$$\begin{aligned} \dot{z} &= F^{-1}AF.z + F^{-1}B \operatorname{diag} \delta_r . w \\ &= K^{-1}E^{-1}AEK.z + K^{-1}E^{-1}B \operatorname{diag} \delta_r . w \\ &= K^{-1}\Lambda K.z + K^{-1}E^{-1}B \operatorname{diag} \delta_r . w \\ &\equiv L.z + M.w \end{aligned} \tag{12.17}$$

where, if λ_c and λ_{c+1} are conjugate complex eigenvalues of A, then

(i) L is formed from diag λ_r by replacing $\begin{bmatrix} \lambda_c & 0 \\ 0 & \lambda_{c+1} \end{bmatrix}$ by $\begin{bmatrix} \Re\lambda_c & \Im\lambda_c \\ -\Im\lambda_c & \Re\lambda_c \end{bmatrix}$

(ii) $M = F^{-1}B \operatorname{diag} \delta_r$, in which F is given by E but with $\Re e_c, \Im e_c$ replacing e_c and e_{c+1} respectively.

Moreover

$$w(t) = \operatorname{diag}(1/\delta_r)\{u(t) - u_{mean}\}$$

and is constrained by

$$1 \geqslant w(t) \geqslant -1$$

while

$$\begin{aligned} z &= F^{-1}X \\ &= F^{-1}(x + A^{-1}Bu_{mean}) \end{aligned} \tag{12.18}$$

Within the framework of this modified system equation, (eqn. 12.17), and the input constraint, (eqn. 12.14), we shall now examine the problem of transferring the state-vector z from some given value z_0 at $t = t_0$ to another given value z_1 at $t = t_1$ in minimum time $(t_1 - t_0)$. As is already known, this problem requires a bang-bang input strategy.

12.5.1 Analysis

Using the H-minimisation technique, the value of H for this problem is $H = 1 + \boldsymbol{k}'\boldsymbol{L}\boldsymbol{z} + \boldsymbol{k}'\boldsymbol{M}\boldsymbol{w}$ and is to be minimised at every instant by a suitable choice of $\boldsymbol{w}$. Since the coefficient of w_r in H is $\boldsymbol{k}'\boldsymbol{m}_r$ (where $\boldsymbol{m}_r$ is the rth column of $\boldsymbol{M}$), w_r is to be switched from + 1 to − 1 (or from − 1 to + 1) whenever $\boldsymbol{k}'\boldsymbol{m}_r$ goes through zero from negative to positive (or from positive to negative). Note that, since $\boldsymbol{w}$ is therefore discontinuous at these switching instants, $\dot{\boldsymbol{z}}$ is discontinuous (although $\boldsymbol{z}$ is not; for in eqn. 12.17 if $\boldsymbol{z}$ were discontinuous, $\dot{\boldsymbol{z}}$ would contain unbalanced impulse-functions) the trajectory in $\boldsymbol{z}$-space is therefore continuous though its direction is in general discontinuous at the switching instants.

Between any two consecutive switching instants, $\boldsymbol{w}$ is constant and its elements have the value + 1 or − 1. Thus during such a stage, assumed for convenience to begin at $t = 0$, we deduce

$$\boldsymbol{z}(t) = \exp(\boldsymbol{L}t).\boldsymbol{z}(0) + \int_0^t \exp\{\boldsymbol{L}(t-\tau)\}.\boldsymbol{M}.\boldsymbol{w}.d\tau$$

$$= \exp(\boldsymbol{L}t).\boldsymbol{z}(0) + \{\exp(\boldsymbol{L}t) - \boldsymbol{I}\}\boldsymbol{L}^{-1}\boldsymbol{M}.\boldsymbol{w}$$

or

$$\boldsymbol{z}(t) + \boldsymbol{L}^{-1}\boldsymbol{M}.\boldsymbol{w} = \exp(\boldsymbol{L}t)\{\boldsymbol{z}(0) + \boldsymbol{L}^{-1}\boldsymbol{M}.\boldsymbol{w}\}$$

or

$$\{\boldsymbol{z}(t) - \boldsymbol{Z}\} = \exp(\boldsymbol{L}t)\{\boldsymbol{z}(0) - \boldsymbol{Z}\} \tag{12.19}$$

where

$$\boldsymbol{Z} \equiv -\boldsymbol{L}^{-1}\boldsymbol{M}.\boldsymbol{w} \tag{12.20}$$

and is calculable for any given $\boldsymbol{w}$.

12.5.1.1 Real eigenvalues of A

Let λ_s be a real eigenvalue of $\boldsymbol{A}$. Then the sth equation of eqn. 12.19 may be written

$$z_s(t) - Z_s = \exp(\lambda_s t)\{z_s(0) - Z_s\} \tag{12.21}$$

where, from eqn. 12.20,

$$-Z_s = (1/\lambda_s)(s\text{th row of } \boldsymbol{M}).\boldsymbol{w} \tag{12.22}$$

Hence, during the control stage considered, $\{z_s(t) - Z_s\}$ cannot change sign. Moreover, if λ_s is negative, $[z_s(t) - Z_s]$ decreases monotonically and asymptotically to zero as $t \to \infty$, but increases monotonically towards infinity if λ_s is positive. Also Z_s is determined by $\boldsymbol{w}$ through eqn. 12.22 and $\boldsymbol{w}$ has 2^m possible values, since each of its m elements may have the value + 1 or − 1; one (or more) of these values of $\boldsymbol{w}$ must determine a maximum positive value $\hat{Z}_s$ of Z_s and its negative then

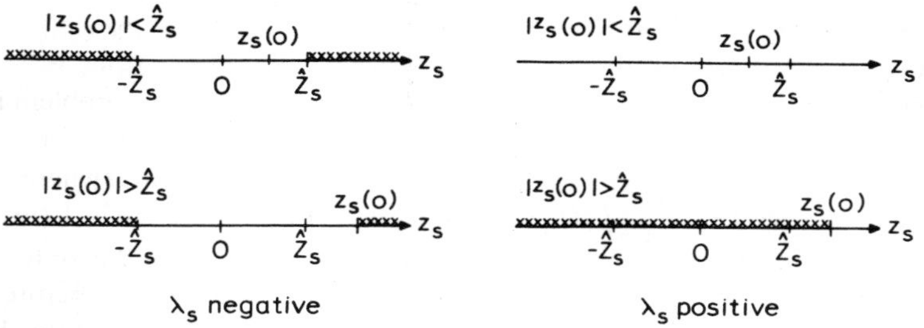

Fig. 12.4 Non-reachable ranges of $z_s(t)$ are shown shaded

determines a negative minimum $-\hat{Z}_s$. It follows that if λ_s is negative and if $z_s(t_0)$ lies between $+Z_s$ and $-Z_s$, $z_s(t)$ must remain within this band during the first control stage and, since it is continuous, during the next control stage etc. and hence during the whole control period. But if $z_s(t_0)$ lies outside this band, it is quite possible to bring its subsequent value $z_s(t)$ to any value between its initial value and the more remote boundary of the band. If on the other hand λ_s is positive and $z_s(t_0)$ lies outside the band $\pm Z_s$, then all the subsequent values of $z_s(t)$ must be of the same sign as the intial value but increase monotonically in modulus, whereas if $z_s(t_0)$ lies inside the band $\pm Z_s$, *any* value of $z_s(t)$ may be reached subsequently by a suitable choice of the w-sequence. These results are illustrated in Fig. 12.4.

Note also that if λ_s, positive or negative, corresponds to a *non-controllable mode* exp $(\lambda_s t)$, then the sth row of $E^{-1}B$ is null, so therefore is the sth row of $E^{-1}B$. diag δ_r, so therefore is the sth row of M, so therefore is Z_s: in this case, clearly, the reachable range of $z_s(t)$ lies between its initial value and zero ($\lambda_s < 0$) or infinity ($\lambda_s > 0$) and moreover its time-behaviour is unaffected by w.

Again, if $\lambda_s \to 0$, then eqn. 12.22 shows that $Z_s \to \infty$; so therefore does $\hat{Z}_s$. In this case, therefore, $z_s(t_0)$ lies within the band $\pm Z_s$ and hence *any* subsequent value of $z_s(t)$ is reachable. In the examples solved earlier, $A = \begin{bmatrix} 0 & 1 \\ 0 & 0 \end{bmatrix}$ and has both eigenvalues zero, which is why, in these problems, any point in the state-plane is reachable from any other point: this is not a typical situation.

All these conclusions are valid for any real eigenvalue of A, whatever the nature of the other eigenvalues. The trajectory in the 1-dimensional z_s-space may be considered as the projection of the trajectory of $z(t)$ in its n-dimensional state-space upon the axis of z_s. As an extension, if there are, say, *two* real eigenvalues, λ_s and λ_p, then from eqn. 12.21, the part-trajectory corresponding to any stage is obtainable by eliminating t between the equation for $z_s(t)$ and that for $z_p(t)$, giving the equation

$$\frac{z_s(t)-Z_s}{z_s(0)-Z_s} = \left\{\frac{z_p(t)-Z_p}{z_p(0)-Z_p}\right\}^{\lambda_s/\lambda_p}$$

This shows that, referred to the point (Z_p, Z_s) as origin, this 2-dimensional trajectory (the projection of the z-space trajectory upon the plane of z_p and z_s) is a simple power-law relationship, the power being positive or negative depending on whether λ_p and λ_s have like or unlike signs.

12.5.1.2 Complex eigenvalues of A

For notational simplicity suppose that $\lambda_1, \lambda_2 = a \pm jb$. Due to the structure of $\boldsymbol{L}$, these eigenvalues will be associated with the principal submatrix $\begin{bmatrix} a & b \\ -b & a \end{bmatrix}$ in $\boldsymbol{L}$ and it is easily shown that the corresponding submatrix in exp $(\boldsymbol{L}t)$ is exp $(at)\begin{bmatrix} \cos bt & \sin bt \\ -\sin bt & \cos bt \end{bmatrix}$. Hence the first two equations of eqn. 12.19 give

$$z_1(t)-Z_1 = \exp at.[\{z_1(0)-Z_1\}\cos bt + \{z_2(0)-Z_2\}\sin bt]$$

$$z_2(t)-Z_2 = \exp at.[-\{z_1(0)-Z_1\}\sin bt + \{z_2(0)-Z_2\}\cos bt]$$

The interpretation of these equations is not so simple as in the case of real eigenvalues. It is clear however that if, for instance, $a < 0$, then $z_1(t) \to Z_1, z_2(t) \to Z_2$, as t increases, but the approach is no longer monotonic but oscillatory; there are still calculable maximum and minimum values for Z_1 and Z_2 etc. Probably the most elegant way of interpreting the above equations is, however, in polar form; for, referred to the point (Z_1, Z_2) as origin the trajectory is easily shown to have the polar equation

$$r = r(0).\exp at = r(0).\exp\frac{a}{b}\{\theta(0)-\theta\}$$

where, moreover, $\dot{\theta} = -b$. Thus although, with $a < 0$, the approach to the point (Z_1, Z_2) may be oscillatory, yet the *radial* approach to this point (or the radial recession *from* this point if $a > 0$) is monotonic.

12.5.1.3 Conclusions

We shall not pursue the analysis further. The main point of carrying it out was to demonstrate that, when the input is constrained in magnitude, there are in general regions of the state space which are *not* reachable from other regions of the state space. (Note in this connection that any region in the z-space maps uniquely, through eqn. 12.18, into some region of the original $\boldsymbol{x}$-space.) It may save considerable time, before trying to minimise the time taken to change $\boldsymbol{x}(t)$ from $\boldsymbol{x}_0$ to $\boldsymbol{x}_1$, to investigate whether such a transfer is possible at all under the input constraints of the problem! Clearly the general conclusions reached in this section with regard to time-minimisation are also applicable to the problem of fuel-minimisation, the main distinction being that, since the control strategy (in terms of $\boldsymbol{u}$) is of the bang-O-bang type, the elements of $\boldsymbol{w}$ are not confined in value to ± 1 but may also adopt some calculable value between these extremes.

12.6 Examples 12

1. The circuit shown is inert at $t = 0$. Determine the control strategy to be imposed on the input voltages u_1, u_2, constrained by $0 \leqslant (u_1, u_2) \leqslant 2$, in order that the capacitor voltages v_1, v_2, may be raised to unity in minimum time and maintained at unity thereafter.

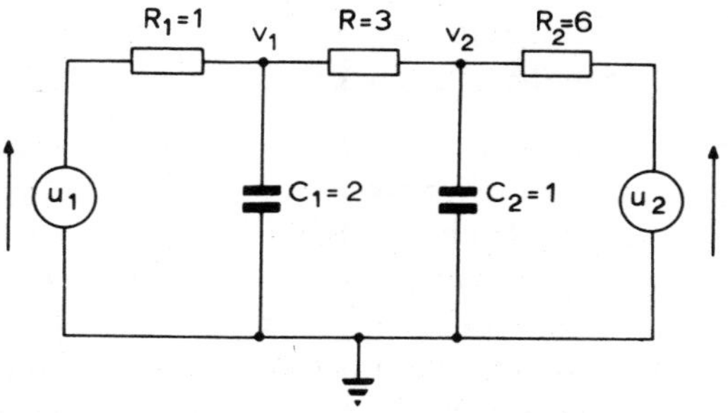

Fig. 12.5 Example 12.1

Treating $\boldsymbol{v} = (v_1, v_2)'$ as a state-vector, obtain the system state-equation and show, from the co-state equation, that the optimal solution contains at most *one* switching operation. Suggest, by logical reasoning, which two input-voltage conditions are to be used in the present problem and in what sequence. Obtain the switching time and the minimum control time to reach the desired conditions. (Note: the internal impedance of the voltage sources is to be taken as zero for all values of u_1 and u_2.)

2. A simple system is defined in normalised form by the equation $\ddot{x} + \dot{x} = u(t)$ where $-1 \leqslant u(t) \leqslant +1$. Obtain a state-vector form of

this differential equation, taking $x_2 = \dot{x}_1$. Deduce the equations of possible trajectories in the (x_1, x_2) plane if u assumes one or other of its extreme values. Obtain also an expression for the time taken between any two points on any such trajectory.

Show from a sketch of these trajectories that, whatever the starting point $\boldsymbol{x}(0)$, the origin may be reached in minimum time with at most one switching operation on u. Evaluate the minimum time required to reach the origin from the point $(-2, 2)$. Can this point be reached from the origin, and if so in what minimum time?

3. For nominal horizontal flight, the control characteristics of a missile can be approximated to by the differential equation $T_m\ddot{c} + \dot{c} = K_1\theta$, where θ is the control-surface deflection and c is the angular heading of the missile. The guidance system of the missile develops a control effort $u(t) = K_2\theta(t)$ and $\int u^2 dt$ is a measure of the fuel used. Determine the *form* of the control signal required to correct any departure from horizontal flight within a fixed control-time T and subject to a minimum-fuel constraint. Obtain expressions for any arbitrary constants in this control form in terms of the control-time T and the initial values of c and $\dot{c}$.

4. A linearised system obeys the state equation $\dot{\boldsymbol{x}} = \begin{bmatrix} 1 & -1 \\ 6 & -4 \end{bmatrix} \boldsymbol{x} + \begin{bmatrix} 8 & 3 \\ 22 & 7 \end{bmatrix} \boldsymbol{u}$ subject to the input constraints $0 \leqslant u_1 \leqslant 1, 0 \leqslant u_2 \leqslant 2$. Convert this equation to the form $\dot{\boldsymbol{z}} = \text{diag}\, \lambda_i . \boldsymbol{z} + \boldsymbol{\beta v}$, where λ_1, λ_2, are the eigenvalues of the plant matrix and $\lambda_1 > \lambda_2$, and where $\boldsymbol{v}$ is a modified input vector such that $-1 \leqslant (v_1 \text{ or } v_2) \leqslant +1$. Express $\boldsymbol{z}$ in terms of $\boldsymbol{x}$ and deduce the value $\boldsymbol{z}_0$ of $\boldsymbol{z}$ corresponding to the origin of $\boldsymbol{x}$. Obtain as much information as possible from the co-state equation and, using this information, sketch a graph of the $\boldsymbol{z}$-plane showing the regions which are reachable from $\boldsymbol{z} = \boldsymbol{z}_0$.

Find the minimum time and the control strategy required to take the system from $\boldsymbol{z} = \boldsymbol{z}_0$ to $\boldsymbol{z} = \begin{bmatrix} 1 \\ -1 \end{bmatrix}$ and show that it is impossible, in the $\boldsymbol{x}$-plane, to reach $\boldsymbol{x} = \begin{bmatrix} 2 \\ 4 \end{bmatrix}$ from the origin.

5. In Example 1 it is required to reach the same capacitor voltages of unity but with minimum energy output from the two voltage sources. The control time, however, must not exceed 6 s. Find the control strategy required and the total energy expended.

Chapter 13

Elements of dynamic programming

13.1 Introduction

The theory of dynamic programming was introduced by Bellman in the 1950s. Although, as the title implies, the theory was primarily developed for the solution of certain problems by digital computer (which implies discrete-time data), it may be extended to continuous-time analysis. The subject is here introduced along the lines of Bellman and Kalaba,[1] an approach which is built up on already familiar concepts.

We begin, roughly in accordance with Chapter 1, with a 2-part definition of a *system*:

(i) a state-vector, symbolised as before by $\boldsymbol{x}(t)$, made up of elements which quantify those aspects of the system which are measurable (the outputs) together with such further elements as may be necessary to ensure that, given $\boldsymbol{x}(t)$ at time t_0, it is possible to forecast the value of $\boldsymbol{x}(t)$ at any later time t_1

(ii) a rule for forecasting: this rule may take a variety of mathematical forms, differential equations in continuous time, for instance $\dot{\boldsymbol{x}}(t) = \boldsymbol{f}(\boldsymbol{x}, t)$, difference equations in discrete time, for instance $\boldsymbol{x}_{k+1} = \boldsymbol{f}(\boldsymbol{x}_k, t)$, or various other forms with which we shall not concern ourselves.

Note that a system, as so defined, is a free system, in the sense that it is devoid of any external inputs.

We next wish to define a *process*, which we shall do in discrete time notation. For this purpose we make use of the state-space trajectory concept: $\boldsymbol{x}(t)$, with co-ordinates $x_1(t), x_2(t), \ldots, x_n(t)$, is represented by a point in the n-dimensional state-space which, as time proceeds, describes a trajectory in this space. In conjunction with this geometrical concept we consider an operator T (which at first we assume time-independent) such that, in discrete time notation, successive sampled values of $x(t)$ are related by $\boldsymbol{x}_{k+1} = T(\boldsymbol{x}_k)$. The set of successive vectors $(\boldsymbol{x}_0, \boldsymbol{x}_1, \boldsymbol{x}_2, \ldots)$, giving the history of the state-vector (or of the point representing it in the state-space) is our definition of a *multistage process*. In particular, an *N-stage process* will be represented by, say, the set $(\boldsymbol{x}_0, \boldsymbol{x}_1, \ldots, \boldsymbol{x}_N)$: note that the *first* value of $\boldsymbol{x}$ corresponds to stage *zero*, and that an N-stage process therefore consists of $(N+1)$ values of the state-vector. Also, with our present definition of the operator T, $\boldsymbol{x}_1 = T(\boldsymbol{x}_0)$, $\boldsymbol{x}_2 = T(\boldsymbol{x}_1) = T^2(\boldsymbol{x}_0)$ etc. (where T^2 signifies operating twice with the operator T), so that an N-stage process may also be represented by $[\boldsymbol{x}_0, T(\boldsymbol{x}_0), \ldots, T^N(\boldsymbol{x}_0)]$.

Note in passing that the processes which we are analysing are assumed to be such that the 'future' state may always be determined by the 'present' state and is not affected, in addition, by the 'past' state: such a process is often referred to as a *Markov process*, though this term is more widely used in connection with stochastic processes than in connection with deterministic processes, which, alone, we are dealing with here.

Note also, to avoid confusion, that the symbol $\boldsymbol{x}_k$ denotes the value of the *vector* $\boldsymbol{x}(t)$ at $t = kh$, where h is the sampling interval; on the other hand the symbol $x_r (r = 1, 2, \ldots, n)$ denotes the rth element of the vector $\boldsymbol{x}$. On the rare occasions when we wish to denote the value of x_r at time kh, we shall use the double suffix symbol x_{rk}, the first number in the suffix corresponding to the element number, the second to the sampling time.

To continue: we shall next wish to consider a variety of scalar functions of the above process, such as

$$\sum_{i=0}^{N} g(\boldsymbol{x}_i) \quad \text{or} \quad \prod_{i=0}^{N} g(\boldsymbol{x}_i) \quad \text{or} \quad \max_{0 \leqslant i \leqslant N} g(\boldsymbol{x}_i) \quad \text{etc.} \quad (13.1)$$

The particular form of function to study will of course depend upon the nature of the optimisation problem considered.

Considering for instance the first of these forms, denote

$$\begin{aligned}\Gamma_N(\boldsymbol{x}_0) &\equiv g(\boldsymbol{x}_0) + g(\boldsymbol{x}_1) + g(\boldsymbol{x}_2) + \ldots + g(\boldsymbol{x}_N) \\ &= g(\boldsymbol{x}_0) + g\{T(\boldsymbol{x}_0)\} + g\{T^2(\boldsymbol{x}_0)\} + \ldots + g\{T^N(\boldsymbol{x}_0)\}\end{aligned}$$

For a given operator T, it is clear that $\Gamma_N(\boldsymbol{x}_0)$ is determined entirely, as the notation implies, by the initial state vector $\boldsymbol{x}_0$ and by N. Moreover, the partial sum obtained by omitting the first term in the last series may be written

$$\begin{aligned}&g(\boldsymbol{x}_1) + g\{T(\boldsymbol{x}_1)\} + \ldots + g\{T^{N-1}(\boldsymbol{x}_1)\} \\ &= \Gamma_{N-1}(\boldsymbol{x}_1),\end{aligned}$$

by the definition of the Γ-function,

$$= \Gamma_{N-1}\{T(x_0)\}$$

Hence Γ satisfies the functional recurrence equation

$$\Gamma_N(x_0) = g(x_0) + \Gamma_{N-1}\{T(x_0)\} \quad \text{with} \quad \Gamma_0(x_0) = g(x_0) \quad (13.2)$$

Similarly, if $\Gamma_N(x_0)$ is identified with the second and third functions in expression 13.1, we deduce the recurrence equations

$$\Gamma_N(x_0) = g(x_0).\Gamma_{N-1}\{T(x_0)\}] \quad \text{with} \quad \Gamma_0(x_0) = g(x_0) \qquad (13.3)$$

and

$$\Gamma_N(x_0) = max\,[g(x_0), \Gamma_{N-1}\{T(x_0)\}] \quad \text{with} \quad \Gamma_0(x_0) = g(x_0) \, (13.4)$$

respectively. The proof is left to the reader.

If the operator T is itself time-dependent, these recurrence equations are of course no longer valid. We now have a relation between successive sampled values of the state-vector which may be written

$$x_{k+1} = T_k(x_k)$$

We may consider in this context an also time-varying form of, say, the first function of eqn. 13.1, namely the function $\sum_{i=m}^{m+N} g_i(x_i)$. An N-stage process may be considered as the set of values $(x_m, x_{m+1}, \ldots, x_{m+N})$. Setting $\Gamma_{N,m}(x_m) \equiv g_m(x_m) + g_{m+1}(x_{m+1}) + \ldots + g_{m+N}(x_{m+N})$, we note that this series, deprived of its first term, is identical with the original series but with $(N-1)$ replacing N and $(m+1)$ replacing m. In other words

$$\Gamma_{N,m}(x_m) = g_m(x_m) + \Gamma_{N-1,m+1}(x_{m+1})$$

$$\left.\begin{aligned} &= g_m(x_m) + \Gamma_{N-1,m+1}\{T_m(x_m)\}, \\ \text{with} \quad \Gamma_{0,m}(x_m) &= g_m(x_m) \end{aligned}\right\} \qquad (13.5)$$

which is a generalisation of eqn. 13.2.

We have so far discussed, in discrete-time notation, what we mean by a system and a multistage process and we have obtained recurrence equations relating to the values of certain scalar functions of this process. We still have to introduce the concept of a multistage *decision* process; we shall also attempt to translate all results obtained into the domain of continuous time.

13.1.1 Multistage decision processes

The operator T relating successive values of the state-vector $\boldsymbol{x}$ has been assumed so far to be an innate property of the system which it serves to define. But suppose now that T, which for simplicity we shall again assume to be time-independent, is a function of a vector $\boldsymbol{u}$, the external input-vector, the value of which may at any time be selected arbitrarily from some given set of permissible values. Then we have successive transformations of the form

$$\boldsymbol{x}_1 = T(\boldsymbol{x}_0, \boldsymbol{u}_0); \boldsymbol{x}_2 = T(\boldsymbol{x}_1, \boldsymbol{u}_1); \ldots ; \boldsymbol{x}_{k+1} = T(\boldsymbol{x}_k, \boldsymbol{u}_k); \ldots$$

If we are given some scalar function of the successive values of $\boldsymbol{x}$ and $\boldsymbol{u}$,

$$J = J(\boldsymbol{x}_0, \boldsymbol{x}_1, \boldsymbol{x}_2, \ldots, \boldsymbol{u}_0, \boldsymbol{u}_1, \boldsymbol{u}_2, \ldots)$$

which is variously called the *Cost Function* (if it is to be minimised), the *Return Function* (if it is to be maximised), or more generally the *Criterion Function*, then the general problem of optimisation of J may be seen as making a suitable choice of the vector $\boldsymbol{u}$ at each stage of the process in such a way as to maximise or minimise J. Since at each stage we have to *decide* what value of $\boldsymbol{u}$ should be used, the vector $\boldsymbol{u}$ is often called the *decision vector* and the optimisation process is, legitimately, called a *multistage decision process*, consisting, in the case of an N-stage process, of the set of vectors $(\boldsymbol{x}_0, \boldsymbol{x}_1, \ldots \boldsymbol{x}_N ; \boldsymbol{u}_0, \boldsymbol{u}_1, \ldots \boldsymbol{u}_{N-1})$.

In the most general case, it may be assumed that the decision to use a particular value of $\boldsymbol{u}$, namely $\boldsymbol{u}_k$ at stage k, is determined by the value of the state vector $\boldsymbol{x}$ at stage k (and possibly at earlier stages also) as well as by the value of $\boldsymbol{u}$ at earlier stages, i.e.

$$\boldsymbol{u}_k = \boldsymbol{u}_k(\boldsymbol{x}_k \boldsymbol{x}_{k-1}, \ldots ; \boldsymbol{u}_{k-1}, \boldsymbol{u}_{k-2}, \ldots)$$

which is called the *policy function*. We shall, however, confine our attention to those problems in which the policy function is only a function of the existing state, i.e.

$$\boldsymbol{u}_k = \boldsymbol{u}_k(\boldsymbol{x}_k)$$

This apparently drastic restriction is more apparent than real. In the first place the earlier values of $\boldsymbol{u}$ have, through the transformation mechanism $\boldsymbol{x}_{n+1} = T(\boldsymbol{x}_n, \boldsymbol{u}_n)$, modified the values of $\ldots \boldsymbol{x}_{k-2}, \boldsymbol{x}_{k-1}$, $\boldsymbol{x}_k$, so that any function of these earlier values of $\boldsymbol{u}$ may be considered as functions of the state vector values at and before stage k. In the second place, if we assume $\boldsymbol{u}_k$ is a function not merely of $\boldsymbol{x}_k$ but also of $\boldsymbol{x}_{k-1}, \boldsymbol{x}_{k-2}$, say, we may coin a new state vector $\boldsymbol{y}$, of three times the order of $\boldsymbol{x}$ and such that $\boldsymbol{y}_k = [\boldsymbol{x}'_k, \boldsymbol{x}'_{k-1}, \boldsymbol{x}'_{k-2}]'$ so that

$u_k = u_k(y_k)$. Whatever the justification, this limitation of the policy function to current-state dependence will be found to lead to a solution of the majority of optimisation problems considered so far: in such cases the optimal policy is determined by Bellman's *Principle of Optimality*, usually stated as follows:

13.1.2 The Principle of Optimality

An optimal policy has the property that, whatever the initial state and decision may be, the remaining decisions constitute an optimal policy with respect to the state resulting from the first decision.

In other words, if we have an N-stage process originating in the initial state $\boldsymbol{x}_0$, with an optimal sequence of decisions $(\boldsymbol{u}_0, \boldsymbol{u}_1, \ldots \boldsymbol{u}_{N-1})$, then the sequence of decisions $(\boldsymbol{u}_1, \boldsymbol{u}_2, \ldots \boldsymbol{u}_{N-1})$ is optimal with respect to the $(N-1)$ stage process originating in $\boldsymbol{x}_1$. To illustrate this fairly common-sense yet vital principle, denote the optimum value of J (maximum or minimum), depending only on $\boldsymbol{x}_0$ and N, by $\hat{J}_N(\boldsymbol{x}_0)$; suppose moreover that $J = \sum_{i=0}^{N} g(\boldsymbol{x}_i, \boldsymbol{u}_i)$. Then the principle of optimality leads to the conclusion that, whatever $\boldsymbol{u}_0$ may have been,

$$\left.\begin{aligned}
\hat{J}_N(\boldsymbol{x}_0) &= \underset{\boldsymbol{u}_0, \boldsymbol{u}_1, \ldots \boldsymbol{u}_N}{\text{Opt}} \{g(\boldsymbol{x}_0, \boldsymbol{u}_0) + g(\boldsymbol{x}_1, \boldsymbol{u}_1) + \ldots + g(\boldsymbol{x}_N, \boldsymbol{u}_N)\} \\
&= \underset{\boldsymbol{u}_0}{\text{Opt}} [g(\boldsymbol{x}_0, \boldsymbol{u}_0) + \underset{\boldsymbol{u}_1, \ldots \boldsymbol{u}_N}{\text{Opt}} \{g(\boldsymbol{x}_1, \boldsymbol{u}_1) + \ldots + g(\boldsymbol{x}_N, \boldsymbol{u}_N)\}] \\
&= \underset{\boldsymbol{u}_0}{\text{Opt}} \{g(\boldsymbol{x}_0, \boldsymbol{u}_0) + \hat{J}_{N-1}(\boldsymbol{x}_1)\} \\
&= \underset{\boldsymbol{u}_0}{\text{Opt}} [g(\boldsymbol{x}_0, \boldsymbol{u}_0) + \hat{J}_{N-1}\{T(\boldsymbol{x}_0, \boldsymbol{u}_0)\}] \\
\text{with} \quad & \\
\hat{J}_0(\boldsymbol{x}_0) &= \underset{\boldsymbol{u}_0}{\text{Opt}}\, g(\boldsymbol{x}_0, \boldsymbol{u}_0)
\end{aligned}\right\} \quad (13.6)$$

In practice, it will be found useful to generalise this recurrence relationship to the final n-stage process of the above N-stage process, of which the initial state is therefore $\boldsymbol{x}_{N-n}$. With obvious modifications we deduce

$$\left.\begin{aligned}
\hat{J}_n(\boldsymbol{x}_{N-n}) &= \underset{\boldsymbol{u}_{N-n}}{\text{Opt}} [g(\boldsymbol{x}_{N-n}, \boldsymbol{u}_{N-n}) + \hat{J}_{n-1}\{T(\boldsymbol{x}_{N-n}, \boldsymbol{u}_{N-n})\}] \\
\text{with} \quad & \\
\hat{J}_0(\boldsymbol{x}_N) &= \underset{\boldsymbol{u}_N}{\text{Opt}}\, g(\boldsymbol{x}_N, \boldsymbol{u}_N) \text{ corresponding to } n = 0.
\end{aligned}\right\} \quad (13.7)$$

13.1.3 Multistage decision process in continuous time

Let the sampling interval in the above discrete-time analysis be h and let $Nh = H$, the total process time. As $h \to 0$, clearly $N \to \infty$ and we shall be dealing with an infinite-stage process: this may, in certain cases, raise problems regarding the existence of limits, but we shall proceed formally.

Consider the criterion function to be of integral form: $J = \int_0^H g\{\boldsymbol{x}(t), \boldsymbol{u}(t)\}\,dt$ a form derivable from the J of the last section by multiplying by h and then making $h \to 0$. We shall also suppose that the transformation T, as $h \to 0$, takes the form $\boldsymbol{x}_{k+1} = T(\boldsymbol{x}_k, \boldsymbol{u}_k) = \boldsymbol{x}_k + h \,.\, \boldsymbol{f}(\boldsymbol{x}_k, \boldsymbol{u}_k)$ giving, in the limit, $\dot{\boldsymbol{x}}(t) = \boldsymbol{f}\{\boldsymbol{x}(t), \boldsymbol{u}(t)\}$, the familiar state equation for a nonlinear, autonomous system. Consider the range of time-integration $0 \leqslant t \leqslant H$ in J to be broken up into the ultimately infinitesimal range $0 \leqslant t \leqslant h$ and the finite range $h \leqslant t \leqslant H$. Define the optimal value of J (which, as before, depends only upon the initial state and the process time) by $\hat{J}[\boldsymbol{x}(0), H]$. Then the principle of optimality gives

$$\hat{J}\{\boldsymbol{x}(0), H\} = \underset{\boldsymbol{u}(t)}{\text{Opt}}\left[\int_0^h g\{\boldsymbol{x}(t), \boldsymbol{u}(t)\}dt + \int_h^H g\{\boldsymbol{x}(t), \boldsymbol{u}(t)\}dt\right]$$
$$= \underset{\boldsymbol{u}(t)}{\text{Opt}}\,(h \,.\, g\{\boldsymbol{x}(0), \boldsymbol{u}(0)\} + \hat{J}[\boldsymbol{x}(0) + hf\{\boldsymbol{x}(0), \boldsymbol{u}(0)\}, H-h])$$

[since, for the second integral, initial state is $\boldsymbol{x}(h) = \boldsymbol{x}(0) + h \,.\, \dot{\boldsymbol{x}}(0)$]

$$= \underset{\boldsymbol{u}(0)}{\text{Opt}}\left[h \,.\, g\{\boldsymbol{x}(0), \boldsymbol{u}(0)\} + \hat{J}\{\boldsymbol{x}(0), H\} + \frac{\partial \hat{J}}{\partial \boldsymbol{x}(0)} \,.\, h \,.\, \boldsymbol{f}\{\boldsymbol{x}(0), \boldsymbol{u}(0)\} - h\frac{\partial \hat{J}}{\partial H}\right]$$

(assuming a Taylor expansion of $\hat{J}$ neglecting higher powers of h)

Note that the optimisation is now with respect to $\boldsymbol{u}(0)$ since this is the only input term in the function to be optimised. Cancelling the term $\hat{J}\{x(0), H\}$, dividing by h and rearranging leads to

$$\left.\begin{aligned} \frac{\partial}{\partial H}\hat{J}\{\boldsymbol{x}(0), H\} &= \left[\underset{\boldsymbol{u}(0)}{\text{Opt}}\;\; g\{\boldsymbol{x}(0), \boldsymbol{u}(0)\} + \frac{\partial \hat{J}}{\partial \boldsymbol{x}(0)} \,.\, \boldsymbol{f}\{\boldsymbol{x}(0), \boldsymbol{u}(0)\}\right] \\ &\text{with} \\ \hat{J}\{\boldsymbol{x}(0), 0\} &= 0 \end{aligned}\right\} \quad (13.8)$$

This equation may be generalised in the same way as eqn. 13.6, this time by considering the process time $t \leqslant \tau \leqslant H$, broken up into the two ranges $t \leqslant \tau \leqslant t + h$ (where $h \to 0$) and $t + h \leqslant \tau \leqslant H$. Noting that initial conditions are now $\boldsymbol{x}(t), \boldsymbol{u}(t)$, and that $(H-t)$ now replaces H so that, considering H fixed, $-\frac{\partial}{\partial t}$ replaces $\frac{\partial}{\partial H}$, we deduce

$$\left.\begin{aligned} &-\frac{\partial}{\partial t}\hat{J}\{\boldsymbol{x}(t), H-t\} \\ &= \underset{\boldsymbol{u}(t)}{\text{Opt}}\left[g\{\boldsymbol{x}(t), \boldsymbol{u}(t)\} + \frac{\partial}{\partial \boldsymbol{x}(t)}\hat{J}\{\boldsymbol{x}(t), H-t\} . \boldsymbol{f}\{\boldsymbol{x}(t), \boldsymbol{u}(t)\}\right] \\ &\text{with} \\ &\hat{J}\{\boldsymbol{x}(H), 0\} = 0 \end{aligned}\right\} \quad (13.9)$$

Eqns. 13.6 to 13.9, particularly the more general forms 13.7 and 13.9, are basic to the theory of dynamic programming. Without too much difficulty they may be extended to time-dependent systems. While we shall return to these basic equations later, we shall first consider a *very* simple example to illustrate the general principles.

13.2 A simple illustrative example

Example: $x(t)$ is a scalar, measurable at $t = 0, h, 2h, \ldots, Nh = H$. Given $x(0) = x_0$ it is required to find the trajectory in the (x, t) plane, i.e. the process $(x_0, x_1, \ldots x_N)$, which will minimise the criterion function $J = \sum_{i=0}^{N} (x_i^2 + u_i^2)$ where u_i is the slope of the line joining (x_i, ih) to $[x_{i+1}, (i+1)h]$. The terminal point x_N is free.

Solution: For this example, $g(x, u) = x^2 + u^2$ and, since $x_{i+1} = x_i + hu_i = T(x_i, u_i)$, it follows that $f(x_i, u_i) = hu_i$. Substituting in eqn. 13.7 gives

$$\hat{J}_n(x_{N-n}) = \underset{u_{N-n}}{\text{Opt}} \{x_{N-n}^2 + u_{N-n}^2 + \hat{J}_{n-1}(x_{N-n} + hu_{N-n})\} \quad (13.10)$$

with

$$\hat{J}_0(x_N) = \underset{u_N}{\text{Opt}} (x_N^2 + u_N^2) = x_N^2 \text{ when } u_N = 0 \text{ (clearly a minimum).}$$

Putting $n = 1$ in eqn. 13.10 and substituting this value of $\hat{J}_0$ gives

$$\hat{J}_1(x_{N-1}) = \underset{u_{N-1}}{\text{Opt}} \{x_{N-1}^2 + u_{N-1}^2 + (x_{N-1} + hu_{N-1})^2\}$$

Differentiating with respect to u_{N-1} and equating to zero (which again clearly gives a minimum) leads to

$$u_{N-1} = -hx_{N-1}/(1+h^2)$$

so that

$$x_N = x_{N-1} + hu_{N-1} = x_{N-1}/(1+h^2) \qquad (13.11)$$

On substitution of these values we find

$$\hat{J}_1(x_{N-1}) = \frac{2+h^2}{1+h^2} x_{N-1}^2$$

We may now substitute this value of $\hat{J}_1$ in eqn. 13.10 with $n = 2$ and, after again differentiating to find the minimum, derive $\hat{J}_2(x_{N-2})$; then $\hat{J}_3(x_{N-3})$ etc. and finally $\hat{J}_N(x_0)$, whatever value N may have. As the conditions for successive minima are found, equations similar to eqn. 13.11 will be obtained, the first part of which relates the input to the state at a particular stage while the second part gives the ratio of successive state values and, since x_0 is given, all other state values may be found. Finally $J_{min} = \hat{J}_N(x_0)$ and $\hat{J}_N$ has been found.

As seen above, $\hat{J}_0(x_N) = x_N^2$ and $\hat{J}_1(x_{N-1}) = \dfrac{2+h^2}{1+h^2} x_{N-1}^2$. This suggests the general rule $\hat{J}_n(x_{N-n}) = K_n x_{N-n}^2$ where K_n is a function of h of which the form depends upon n. On substituting for $\hat{J}_n$ and $\hat{J}_{n-1}$ in eqn. 13.10 we deduce:

$$K_n x_{N-n}^2 = \min_{u_{N-n}} \{x_{N-n}^2 + u_{N-n}^2 + K_{n-1}(x_{N-n} + hu_{N-n})^2\}$$

On performing the minimisation by differentiation, there result

$$u_{N-n} = -hK_{n-1}x_{N-n}/(1+h^2K_{n-1});$$

$$x_{N-n+1} = x_{N-n} + hu_{N-n} = x_{N-n}/(1+h^2K_{n-1}) \qquad (13.12)$$

which, on substitution, give

$$K_n = 1 + K_{n-1}/(1+h^2K_{n-1}) \quad \text{with} \quad K_0 = 1, \qquad (13.13)$$

as already found.

Comments

(a) All the above iteration processes are ideal for computerisation for any given, numerical value of h

(b) From eqn. 13.13, if $K_{n-1} > 0$, so is K_n; but $K_0 = 1$, so that K_1, $K_2, \ldots, K_N > 0$

(c) It then follows from the second part of eqn. 13.12 that x_{N-n+1} and x_{N-n} have the same sign; if $x_0 > 0$, all x_n are therefore positive; moreover from the same equation, x_n decreases monotonically as n increases.

(d) Combining both parts of eqn. 13.12

$$u_{N-n} = -hK_{n-1}x_{N-n+1};$$

hence

$$u_{N-n}/u_{N-n-1} = K_{n-1}x_{N-n+1}/K_n x_{N-n}$$
$$= K_{n-1}/\{1+(1+h^2)K_{n-1}\} < 1$$

Hence u is always of the opposite sign to x, and therefore negative, and moreover its magnitude decreases monotonically towards the terminal value $u_N = 0$ already found.

(e) From the first part of eqn. 13.11, u_{N-n} is proportional to x_{N-n} through the medium of the time-varying factor $-hK_{n-1}/(1+h^2K_{n-1}) = -h(K_n - 1)$. Thus the optimising input u, assumed constant over any one sampling interval, may be derived from the state, x, by a time-varying negative-feedback link.

(f) The calculations only make use of the boundary condition $x(0) = x_0$: the 2-point boundary problem appears to be absent.

To illustrate some of these points consider the numerical case $H = 2$, $N = 6$, so that $h = 1/3$. In this case eqn. 13.13 gives $K_n = \dfrac{9+10K_{n-1}}{9+K_{n-1}}$ and, with $K_0 = 1$, we find successively $K_1 = \dfrac{19}{10}$, $K_2 = \dfrac{280}{109}$, $K_3 = \dfrac{3781}{1261}$, $K_4 = \dfrac{49159}{15130}$, $K_5 = \dfrac{627760}{185329}$, $K_6 = \dfrac{7945561}{2295721}$, and $J_{min} = K_6 x_0^2$. Using these values of K_n in the second part of eqn. 13.12 gives, by iteration,

$$x_0 : x_1 : x_2 : x_3 : x_4 : x_5 : x_6$$
$$= 1 : 0{\cdot}7266 : 0{\cdot}5338 : 0{\cdot}4004 : 0{\cdot}3115 : 0{\cdot}2572 : 0{\cdot}2315$$

and using these values in turn in the relation $x_{n+1} = x_n + hu_n$ leads to

$$x_0 : u_0 : u_1 : u_2 : u_3 : u_4 : u_5 : u_6$$
$$= -1 : 0{\cdot}820 : 0{\cdot}578 : 0{\cdot}400 : 0{\cdot}267 : 0{\cdot}163 : 0{\cdot}077 : 0$$

results which the reader is invited to check!

13.2.1 Optimal path through grid network

We now solve the same problem by another application of the principle of optimality, taking $x_0 = 3$ for instance. We suppose that at any stage of the process, the value of x is constrained to be one of a set of equispaced values: the finer the spacing of this x-grid, the more accurate

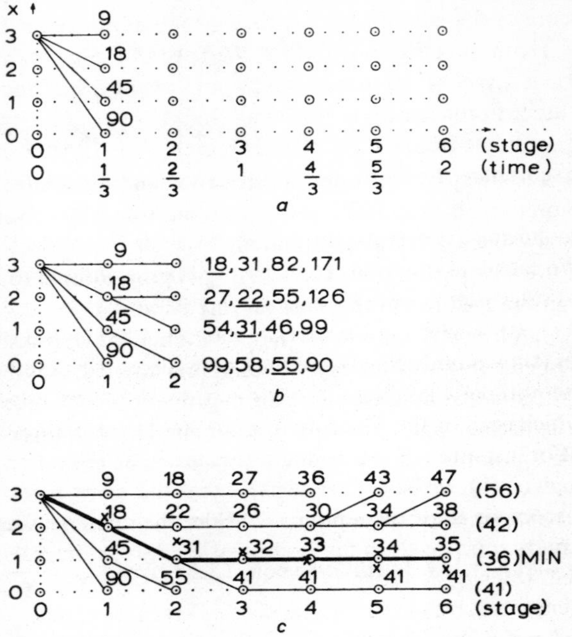

Fig. 13.1 (*a*) The grid and the four paths from stage 0 to stage 1
(*b*) Minimising the score to stage 2. The four figures at each stage 2 point are the total scores obtained by reaching each point from (1, 3), (1, 2), (1, 1), and (1, 0), respectively. Minima are underlined, and minimal paths are shown
(*c*) The diagram completed. The optimal path is reinforced

the answer but the longer the computational process! To illustrate, we shall use a coarse grid of *integer* values of x and we shall assume further that x is confined to the four values $x = 0, 1, 2, 3$: we are of course guided in this choice by the value of x_0 and by the analysis of the last section which shows that x is positive but decreases with time, but, putting the argument in reverse, the present method is *expedited* by placing range restrictions on x (in general, on the state vector), whereas such limits seriously complicate all previous optimisation techniques considered.

The stage-state grid is shown in Fig. 13.1*a*. The location of any point on the grid will be denoted by (a, b) where a = stage-number, b = value of x. Thus we start from (0, 3). In going from one stage to the next there are four possible paths to follow from any given starting point. Since u is the slope and the stage-time is 1/3, it follows that

$u = 3$ (increase in x between stages). As we travel we shall keep a running score of the quantity we are trying to minimise, namely $\Sigma(x_i^2 + u_i^2)$, counting the contribution made to this quantity during any interstage travel as the sum of the squares of the *initial* state and the slope used. For instance, in travelling from (0, 3) to (1, 3) the score is $3^2 + 0^2 = 9$; but from (0, 3) to (1, 2) scores $3^2 + 3^2 = 18$ etc. These first stage scores are shown in the figure: in computer work they must be stored.

The second stage is more arduous since there are 16 paths to be considered! We calculate the total score achieved in reaching any one of the second-stage points from each of the first-stage points (1, 3), (1, 2), (1, 1) and (1, 0), as given in Fig. 13.1*b*, and select the lowest of these four scores (shown underlined): we may now forget the other scores BUT, if the optimal x and u are to be found, we must remember the path by which each of the four minimum scores at stage 2 has been achieved. For instance, IF it is found ultimately that the optimal path goes through (2, 1), we have to remember that this point is reached with minimum score via the x-sequence (3, 2, 1) as shown in the diagram. The construction proceeds in the same way for further stages until stage 6 is reached, when x_6^2 is added to each score. We argue that u_6, which only influences x_7, is not part of the process considered, but since J contains u_6^2 and has to be minimised, we take $u_6 = 0$ (the same result as by the first method used). The final diagram is Fig. 13.1*c* and shows the minimum score of 36 to be reached by the x-sequence (3, 2, 1, 1, 1, 1, 1) and the u-sequence $(-3, -3, 0, 0, 0, 0, 0)$. These results should be compared with the results of Section 13.2, remembering that $x_3 = 3$. The accuracy is admittedly poor, but this is largely due to the coarse x-grid used, chosen to avoid too much arithmetic while adequately demonstrating the method.

Comments:

(*a*) The problem is *very* simple: the state is scalar, the state equation, $\dot{x} = u$, is simple and linear, the criterion function J is a simple quadratic form and all parameters are time-invariant. It must be appreciated, however, that the method is applicable in the most general case of a multidimensional state vector (though this of course prevents a 2-dimensional diagrammatic representation), nonlinear time-dependent state equations and quite general criterion functions. Indeed if the state equation, in continuous time, is $\dot{\mathbf{x}} = \mathbf{f}(\mathbf{x}, \mathbf{u}, t)$ which, in discrete time, may be approximated by $\mathbf{x}_{k+1} = \mathbf{x}_k + h\mathbf{f}(\mathbf{x}_k, \mathbf{u}_k, hk)$, then provided that it is possible to solve this equation for the elements of $\mathbf{u}_k$ (given h, k and specified grid values of x_k and x_{k+1}), it is then possible to substitute for u_k in any criterion function of the form $J = \sum_k g_k(x_k, u_k, hk)$ to find the contribution made to J by any path linking stage k to stage $(k + 1)$.

(*b*) It has already been noted that the method is expedited by placing constraints on the elements of $\boldsymbol{x}$. The same is true if similar constraints apply to $\boldsymbol{u}$. In the example, if $-3 \leqslant u \leqslant 0$, say, the solution would be unchanged since the optimising u at no stage goes outside these limits: however, the solution would be reached more rapidly because in any interstage displacement, instead of having to consider four possible paths, we need only consider a drop in x of 0 or 1. Thus we have two paths from stage 0 to stage 1 (instead of four), four paths from stage 1 to stage 2 (instead of sixteen) etc.

(*c*) *The method is equally applicable if the final state is given instead of free*. If for instance we are given $x_6 = 2$, the only difference in the solution process would be that in going from stage 5 to stage 6 we need *only* consider the four paths terminating in (6, 2) and the scores for paths terminating in (6, 0), (6, 1) and (6, 3) are irrelevant. $J_{min} = 42$ and $x_r = 2$ $(r = 1, 2, 3, 4, 5, 6)$. See Fig. 13.1*c*. Similarly, if x_6 were given as 0; but it might be wise in this case to extend the x-grid say to (-1) or even (-2) in case the optimal path contains a negative value of x; this grid-extension is usually a wise policy to pursue whenever it is found that the calculated optimal path impinges upon arbitrarily determined grid boundaries.

(*d*) In the absence of the principle of optimality, we could of course, in this example, have computed the values of J obtained by each of the $4^6 = 4096$ paths linking $x_0 = 3$ to $x_6 = 0, 1, 2$ or 3: a rather lengthy operation, involving, moreover, 4095 comparisons to establish the minimum score. The method used has reduced this number of comparisons to three at each of the four grid points of stages 2 to 6, together with a final three to compare the four minimum scores at stage 6, a total of 63: a very worthwhile reduction. But this problem has a 1-dimensional state and a coarse grid. More generally, let the state x be n-dimensional, let the grid allow g possible values *of each element* of x and let there be N-stages (apart from stage 0). There will now be g^n paths linking any given starting-point to possible stage-1 points; for each such path, there will be a further g^n paths to possible stage-2 points etc., making a total of g^{nN} paths, entailing $(g^{nN} - 1) \doteqdot g^{nN}$ score-comparisons if we compute each through-path separately. If, however, we use an extension of the method used above, we have $(g^n - 1)$ comparisons at each of the g^n grid points of stages 2 to N, together with $(g^n - 1)$ final comparisons at stage N, a total of $(g^n - 1)[(N-1)g^n + 1] \doteqdot (N-1)g^{2n}$, giving a reduction ratio of about $(N-1)/g^{n(N-2)}$. If for instance $n = N = 6$ and $g = 10$ the reduction ratio is about $5 \cdot 10^{-24}$ (*very* worthwhile!), BUT the reduced number of comparisons is still some 5×10^{12}, and before we can make any comparisons we have to compute the contribution to the cost-function of some 5×10^{12} interstage paths, quite apart from the memory-loading problem of storing the trajectories which, at any stage-point, gave a minimum score. This illustrates what Bellman aptly called the 'curse of dimensionality', which clearly limits the method, even when using modern computers, to problems involving relatively low values of n.

13.3 Linear autonomous system and quadratic criterion integral

The example of Section 13.2, although presented initially as the problem of finding an optimal trajectory in the (t, x) plane, is clearly the problem of a linear autonomous system, represented by $\dot{x} = u$, with a quadratic form criterion integral $J = \int_0^H (x^2 + u^2)dt$, treated in discrete-time notation. We shall now consider the application of the same techniques to the more general problem of this type, already considered in Chapter 11.

The system state equation is assumed to be

$$\dot{x} = Ax + Bu \text{ in continuous time;}$$

$$x_{k+1} = Fx_k + Gu_k \text{ in discrete time.}$$

The criterion function is assumed to be

$$J = \int_0^H \tfrac{1}{2}(x'Px + u'Qu)dt \text{ (continuous time),}$$

$$J = h \sum_0^N \tfrac{1}{2}(x_k'Px_k + u_k'Qu_k) \text{ (discrete time)}$$

where Q is positive-definite, P is non-negative-definite and $H = Nh$. It is assumed that the initial value of x is given but that the final value is free.

13.3.1 Solution of problem in discrete time

In the present problem $g(x_k, u_k) = \dfrac{h}{2}(x_k'Px_k + u_k'Qu_k)$ and

$x_{k+1} = T(x_k, u_k) = Fx_k + Gu_k$. Substituting in eqn. 13.7 gives:

$$\left.\begin{aligned} \hat{J}_n(x_{N-n}) &= \underset{u_{N-n}}{\text{Min}} \left\{ \frac{h}{2}(x'_{N-n}Px_{N-n} + u'_{N-n}Qu_{N-n}) \right. \\ &\quad \left. + \hat{J}_{n-1}(Fx_{N-n} + Gu_{N-n}) \right\} \\ \text{with} \qquad & \\ \hat{J}_0(x_N) &= \underset{u_N}{\text{Min}} \frac{h}{2}(x'_N Px_N + u'_N Qu_N) \end{aligned}\right\} \quad (13.14)$$

this last minimum being given by $u_N = 0$ since Q is positive definite. Hence $\hat{J}_0(x_N) = \frac{h}{2} x'_N P x_N$. Substituting for $\hat{J}_0$ in eqn. 13.14 with $n = 1$ will show that $\hat{J}_1(x_{N-1})$ is a quadratic form in the elements of x_{N-1}; assuming, in parallel with Section 13.2, that

$$\hat{J}_n(x_{N-n}) = \frac{h}{2} x'_{N-n} K_n x_{N-n}$$

where K_n is now a square matrix which, without loss of generality, we may assume to be symmetric, then substitution in eqn. 13.14 gives

$$\begin{aligned} x'_{N-n}K_n x_{N-n} &= \underset{u_{N-n}}{\text{Min}} \{x'_{N-n}Px_{N-n} + u'_{N-n}Qu_{N-n} \\ &\quad + (Fx_{N-n} + Gu_{N-n})'.K_{n-1}(Fx_{N-n} + Gu_{N-n})\} \end{aligned} \quad (13.15)$$

Differentiating with respect to u_{N-n} leads to

$$2u'_{N-n}Q + 2u'_{N-n}G'K_{n-1}G + 2x'_{N-n}F'K_{n-1}G = 0$$

giving

$$u_{N-n} = -(Q + G'K_{n-1}G)^{-1}G'K_{n-1}Fx_{N-n} \quad (13.16)$$

and

$$Fx_{N-n} + Gu_{N-n} = \{I - G(Q + G'K_{n-1}G)^{-1}G'K_{n-1}\}Fx_{N-n} \quad (13.17)$$

On substituting these expressions in eqn. 13.15 we obtain

$$\begin{aligned} x'_{N-n}K_n x_{N-n} = x'_{N-n}[P &+ \{(Q + G'K_{n-1}G)^{-1}G'K_{n-1}F\}' \\ &\times Q\{(Q + G'K_{n-1}G)^{-1}G'K_{n-1}F\} \\ &+ \{I - G(Q + G'K_{n-1}G)^{-1}G'K_{n-1}F\}' \\ &\times K_{n-1}\{I - G(Q + G'K_{n-1}G)^{-1}G'K_{n-1}F\}]x_{N-n} \end{aligned}$$

If this equality is to hold for all values of x_{N-n} we may suppress the premultiplier x'_{N-n} and the postmultiplier x_{N-n} on both sides and obtain an explicit solution for K_n in terms of K_{n-1}; since by eqn. 13.14

which is the co-state equation of Chapter 11.

We note some further points in connection with the Riccati equation.

13.3.2.1 Comments on the solution of the Riccati equation

(i) If the Riccati equation can be solved, we then have $\dot{\boldsymbol{x}} = \boldsymbol{A}\boldsymbol{x} + \boldsymbol{B}\boldsymbol{u} = \{\boldsymbol{A} - \boldsymbol{S}\boldsymbol{K}(t)\}\boldsymbol{x}$ as an equation for $\boldsymbol{x}$ with the boundary condition $\boldsymbol{x}(0)$ given. After solving this equation for $\boldsymbol{x}$, $\boldsymbol{u}$ may be found from eqn. 13.20. Since in the present problem the boundary condition on $\boldsymbol{K}$ is $\boldsymbol{K}(H) = \boldsymbol{0}$, the solution for $\boldsymbol{K}$ in the nonlinear Riccati equation will normally be carried out by computer, starting at the final time $t = H$ and working backwards in time. The solution for $\boldsymbol{x}$ on the other hand will work forwards in time from $t = 0$.

(ii) Since $\boldsymbol{u} = -\boldsymbol{Q}^{-1}\boldsymbol{B}'\boldsymbol{K}\boldsymbol{x}$, the optimising input may be derived from the state by the (normally time-dependent) feed-back link $-\boldsymbol{Q}^{-1}\boldsymbol{B}'\boldsymbol{K}(t)$. In this connection it has been shown by Kalman that, if $H \to \infty$, the solution for $\boldsymbol{K}$ tends to a constant matrix, which is still a solution of eqn. 13.21 but, of course, with $\dot{\boldsymbol{K}} = \boldsymbol{0}$. This value of $\boldsymbol{K}$ is usually called the ***algebraic*** solution of the Riccati equation. In practice, provided that H is two or three times the largest time constant present in the plant, this constant value of $\boldsymbol{K}$, which is of course simpler to engineer than the optimal time-dependent value, will be found to give almost the same value of J: the advantage of the constant feed-back then normally outweighs the small disadvantage of a slightly sub-optimal control.

(iii) The nonlinearity of the Riccati equation means that, although it is a first-order equation and, at any rate in the present example, we have the boundary condition $\boldsymbol{K}(H) = 0$, yet there may be a multiplicity of solutions. This difficulty may be removed by noting that, since $\boldsymbol{P}$ is non-negative definite and $\boldsymbol{Q}$ is positive definite, J is normally positive; so therefore is its minimum value $\hat{J}$.We may then infer from eqn. 13.19 that $\boldsymbol{K}(t)$ is a positive definite matrix: this constraint on $\boldsymbol{K}(t)$ is found to make the solution of the Riccati equation unique. (This argument also applies to the algebraic solution.)

(iv) If the optimal input $\boldsymbol{u}$ could be externally manufactured and fed into the plant, the stability of the resulting system would be identical with the stability of the plant and determined by the eigenvalues of $\boldsymbol{A}$. If, however, which is more usual, $\boldsymbol{u}$ is derived as a feed-back from the state $\boldsymbol{x}$, then we have a closed-loop system for which $\boldsymbol{x}$, as seen above is controlled by $\dot{\boldsymbol{x}} = \{\boldsymbol{A} - \boldsymbol{S}\boldsymbol{K}(t)\}\boldsymbol{x}$, the stability of which requires investigation. This is most easily done by the state and co-state equations of eqn. 11.21 which show that the elements of both $\boldsymbol{x}$ and $\boldsymbol{k}$ contain the modes dictated by the eigenvalues of the Hamiltonian matrix $\begin{bmatrix} A & -S \\ -P & -A' \end{bmatrix}$ which are known to exist in equal and opposite pairs: thus half these modes are stable and half are unstable, so that the closed-loop system is unstable. These modes are independent of H: but if $H \to \infty$ the *coefficients* of the unstable modes will be found to tend to zero. This is to be expected, for unless this is the case, J will not converge. Since therefore in this case $\boldsymbol{x}$ will consist entirely of stable modes, the resulting closed-loop system is stable, *even if the plant is unstable*. Our conclusions on stability are therefore: if the algebraic value of $\boldsymbol{K}$ is used, the resulting closed-loop system is always stable; if, with H finite, the optimal $\boldsymbol{K}(t)$ is used, the resulting closed-loop system is unstable and may therefore lead to dangerous situations if it is maintained after $t = H$; if, in this last case, the plant is itself unstable, the suppression of the feed-back after $t = H$ will still leave an unstable system and either the plant should then be shut down or it must be stabilised by a suitable feed-back (e.g. the algebraic value of $\boldsymbol{K}$).

13.4 On terminal conditions

It has been assumed, since the introduction of the concept of a multi-stage decision process, that the choice of $\boldsymbol{u}$ at any stage of the process is a free choice. Clearly, however, this is no longer true if at the end of the process, $t = H$, $\boldsymbol{x}$ or any of its elements are given, for in this case the optimum value of $\boldsymbol{u}$ at any stage will be affected by this constraint. We have therefore, throughout this Chapter, assumed that $\boldsymbol{x}(H)$ is free. If this is not the case then some of the theory of this Chapter needs revision in the light of the implied constraint: notably, it is only because $x(H)$ is considered free that we have the boundary condition $\boldsymbol{K}(H) = 0$; with certain types of boundary conditions it may not even be true that $\boldsymbol{K}(t)$ is symmetric. A useful introduction to the use of Dynamic Programming techniques in the presence of constraints, including the type mentioned, is given by Jacobs[2] and in general such constraints make the method much more laborious than the unconstrained theory given so far might lead one to expect. While space considerations forbid a treatment of 'constraint' theory here, we conclude with a simple example in which the terminal value of the state is given instead of being free.

13.4.1 Example with terminal State Constraint

In discrete-time notation, $x_{k+1} = T(x_k, u_k) = x_k + hu_k$. Given x_0 and x_N find $u_0, u_1, \ldots u_N$ to minimise $J = \sum_{k=0}^{N} (x_k^2 + u_k^2)$. (Compare Section 13.2)

Solution: Since $x_{k+1} = x_k + hu_k$, it follows, by adding such equations over the values $k = 0, 1, \ldots, (N-1)$, that $x_N - x_0 = h \sum_{k=0}^{N-1} u_k$, and since x_0, x_N and h are given, an equivalent formulation of the terminal state constraint is

$$u_0 + u_1 + \ldots + u_{N-1} = \frac{1}{h}(x_N - x_0) = \text{given constant},$$

showing that the choice of any particular u_k is not free but constrained by the choice of the others. As far as u_N is concerned, in order to minimise J we make u_N zero since it has no effect on any of the relevant x_k.

In spite of this reformulation of the constraint we shall adhere to its original form: x_N is given. We shall use the transformation formula in reverse, i.e. not in order to deduce x_{k+1} from a given x_k and a given u_k but to deduce $u_k = (x_{k+1} - x_k)/h$ from a given x_k and x_{k+1}. Following the same line of thought, we consider J as being minimised not by a choice of u_k at stage k but by a choice of x_{k+1} at stage k. Moreover these choices, as well as the implied values of u_k and hence the optimal contribution to J over, say, the last n stages, will all depend upon x_N, as well as (as before) on x_0, the initial value of x. We write therefore

$\hat{J}_n(x_{N-n}, x_N)$ = minimum contribution to J from stage $(N-n)$ to stage N with respect to $x_{N-n}, x_{N-n+1}, \ldots x_N$, *with* x_{N-n} *and* x_N *given.*

The prinicple of optimality then leads to

$$\begin{aligned}\hat{J}_n(x_{N-n}, x_N) &= \min_{x_{N-n+1}} \{x_{N-n}^2 + u_{N-n}^2 + \hat{J}_{n-1}(x_{N-n+1}, x_N)\} \\ &= \min_{x_{N-n+1}} \{x_{N-n}^2 + (x_{N-n+1} - x_{N-n})^2/h^2 \\ &\quad + \hat{J}_{n-1}(x_{N-n+1}, x_N)\} \end{aligned} \qquad (13.22)$$

Note that if $n = 1$, no minimisation process is permissible since the minimisation is then with respect to x_N, which is fixed. In fact

$$\hat{J}_1(x_{N-1}, x_N) = \frac{1+h^2}{h^2}(x_{N-1}^2 + x_N^2) - 2x_{N-1}x_N/h^2$$

since

$$\hat{J}_0(x_N, x_N) = 0.$$

On substituting this form of $\hat{J}_1(x_{N-1}, x_N)$ in eqn. 13.22 with $n = 2$ we obtain:

$$\hat{J}_2(x_{N-2}, x_N) = \underset{x_{N-1}}{\text{Min}}\left\{x_{N-2}^2 + (x_{N-1} - x_{N-2})^2/h^2 + \frac{1+h^2}{h^2}(x_{N-1}^2 + x_N^2) - \frac{2}{h^2}x_{N-1}x_N\right\}$$

The minimisation condition gives

$$(x_{N-1} - x_{N-2}) + (1+h^2)x_{N-1} - x_N = 0$$

i.e.

$$x_{N-1} = (x_{N-2} + x_N)/(2+h^2)$$

and, on substitution, this value of x_{N-1} leads to

$$\hat{J}_2(x_{N-2}, x_N) = \{(1 + 3h^2 + h^4)(x_{N-2}^2 + x_N^2) - 2x_{N-2}x_N\}/(2+h^2)h^2$$

Substituting this in eqn. 13.22 with $n = 3$ and minimising gives $\hat{J}_3(x_{N-3}, x_N)$ etc. iteratively until $\hat{J}_N(x_0, x_N)$ is found, which is the value of J_{min}.

Alternatively, prompted by the form of $\hat{J}_1$ and $\hat{J}_2$, we may postulate

$$\hat{J}_n(x_{N-n}, x_N) = \frac{a_n(x_{N-n}^2 + x_N^2) - 2x_{N-n}x_N}{h^2 b_n} \tag{13.23}$$

where a_n, b_n are polynomials in h^2. On substituting this form in eqn. 13.22 and then carrying out the minimisation and finally equating coefficients of x_N^2, x_{N-n}^2 and $x_N x_{N-n}$ on each side, it is found that a_n and b_n must satisfy the *three* equations $a_n = (1+h^2)a_{n-1} + h^2 b_{n-1}$, $b_n = a_{n-1} + b_{n-1}$ and $a_n b_{n-1} - a_{n-1}b_n + 1 = 0$. The third equation is found to be consistent with the first two provided that $a_{n-1}^2 - h^2 b_{n-1}^2 - h^2 a_{n-1}b_{n-1} = 1$ and the same relation then holds between a_n and b_n. Since this consistency relation holds for the values $a_1 = 1 + h^2$, $b_1 = 1$ obtainable from $\hat{J}_1(x_{N-1}, x_N)$ above, it holds for all values of n so that a_n and b_n can be found iteratively from the first two equations, starting with $n = 2$, until a_N, b_N are found: these may then be substituted in eqn. 13.23 to give $\hat{J}_N(x_0, x_N) = J_{min}$.

The minimisation condition is found to be

$$x_{N-n+1} = (b_{n-1}x_{N-n} + x_N)/b_n$$

so that

$$u_{N-n} = (x_{N-n+1} - x_{N-n})/h = (x_N - a_{n-1}x_{N-n})/hb_n \quad (13.24)$$

The verification of all these results is an excellent exercise for the reader!

Comments

(i) Even in this very simple example the treatment has been considerably complicated by the fact that the terminal state is given and not free.

(ii) In spite of this, the treatment still avoids the difficulties of the 2-point boundary situation.

(iii) The expression of eqn. 13.24 for u_{N-n} does *not* show, explicitly, that the required optimal input can be derived as a time-dependent feed-back from the current state: compare eqn. 13.11. However eqn. 13.24 may be re-written as $-u_{N-n}/x_{N-n} = (a_{n-1} - x_N/x_{N-1})/hb_n$, which is of course a time-dependent multiplier.

13.5 References

1 BELLMAN, R.E., and KALABA, R.: 'Dynamic programming and control theory' (Academic Press, 1966)

2 JACOBS, O.L.R.: 'An introduction to dynamic programming' (Chapman & Hall, 1967)

13.6 Examples 13

1. A scalar quantity $x(t)$ obeys the equation $\dot{x} = u - \frac{x^2}{2}$. Express this equation in discrete time. With $H = 3, N = 6, x_0 = 3$ and x_N free, find the trajectory (plotting x_k against k) required to minimise $J = h \sum_0^6 (2|x_k| + |u_k|)$ and the resulting value of J_{min}. Use a grid of integer values of x from -1 to $+3$ inclusive. (It should be possible to forecast these results after computing scores at stage 2.)

Analyse the first sampling interval more closely by repeating the problem but with $H = \frac{1}{2}, N = 5$, (i.e. $h = 0{\cdot}1$) using the same x-grid values.

Can you draw any conclusions as to what value of h will minimise the calculated value of J_{min}? What is this limiting value?

2. With reference to the problem of Section 13.2 and especially to eqn. 13.13, obtain the limit-form of K_N as $h \to 0, N \to \infty$, with $Nh = H =$ constant. Deduce the value of hJ_{min} for the problem treated in continuous time. [Suggestion: Suppose $K_n = (a_{no} + a_{n1}h^2 + \ldots + a_{nn}h^{2n})/(1 + b_{n1}h^2 + \ldots + b_{nn}h^{2n})$ (and use eqn. 13.13 to obtain relations between the a_{nk} and b_{nk} and the $a_{(n-1)k}$ and $b_{(n-1)k}$]

Solve the problem directly in continuous time by, say, the Calculus of Variations, and compare results.

3. In the problem of Section 13.4.1, take $x_0 = 10, x_N = 2, N = 6, H = 2$. Compute $x_1, x_2, \ldots x_5$ and $u_0, u, \ldots u_5$. Solve the problem in continuous time, using the Calculus of Variations or any any other method and compare the resulting values of $x(t)$ and $u(t)$. Compare also the values of J_{min}.

4. Find J_{min} in Example 3 by the grid method, with the added constraint that $0 \geqslant u \geqslant -6$. Use a grid of integer values of x from 0 to 10 inclusive.

Chapter 14

On hill climbing

14.1 Introduction

The last four Chapters have been devoted to a particular type of optimisation problem in which it is required to find an input vector, $\boldsymbol{u}(t)$, such that some criterion function, J, dependent in some way on $\boldsymbol{u}(t)$ and usually of integral form, is maximised or minimised. In this Chapter we are concerned with a different type of optimisation problem in which the criterion function – to be minimised or maximised – depends for its value upon certain parameters of the system (*not* normally functions of time), for instance an adjustable rate of flow in a chemical process, the temperature at which the chemical reaction is conducted, the gain of an amplifier, the petrol-to-air ratio feeding an internal combustion engine etc. (Note in passing that, as far as inputs are concerned, we may still include them under the heading of parameters, provided that we first express them, for instance, as a power series in time t, e.g. $u(t) = a_0 + a_1 t + a_2 t^2 + \ldots$, or as a sum of exponentials, e.g. $u(t) = b_0 + b_1 \exp(c_1 t) + b_2 \exp(c_2 t) + \ldots$, or in terms of some other suitable form of time-function; the a_r, the b_r, the c_r in these expressions are all, in the present context, adjustable parameters).

The criterion function J is, like the parameters, not usually a function of time: it may be, for instance, the (steady-state) rate of yield of a chemical process, the (steady-state) efficiency of a boiler etc. It is assumed that J is nonlinear in the parameters considered. Moreover there *need* be no mathematical model of the plant: in the absence of such a model, whether linear or nonlinear, J must be obtained directly or indirectly from experimental measurements on the system, measurements which (a) may be contaminated by 'noise' or other disturbances and (b) may take time to obtain since we are interested in steady-state values. If a mathematical model of some form is available, the value of

J may be computed from some expression giving its steady-state value. These measurements or computations of J are to be made for different sets of values of the various adjustable parameters: we are therefore concerned with discrete values of J estimated in some way at discrete points in the parameter-space.

It may reasonably be suggested that since this problem (in the presence of a mathematical model) is the finding of a minimum or maximum of a function of several parameter-variables, its solution may be obtained by the normal technique of equating every partial derivative of J to zero and determining the nature of the turning value so found by considering the second partial derivatives. This is true. But it must be realised that if J is a highly nonlinear function of, say, 20 parameters, the turning-value equations require the solution of 20 nonlinear, simultaneous equations for 20 variables! (We omit the findings of 190 second derivatives!) Moreover these equations, being nonlinear, will normally have multiple solutions, each of which must be investigated to determine whether it is in fact the true maximum or minimum required.

The methods outlined in this Chapter avoid the complications of this analytical process and are all based upon the assessment of J at various points in the parameter space, points so chosen that an approximation to the point where J is maximised or minimised may be achieved as quickly as possible, which means, whether J is assessed experimentally or computationally, with as few trials as possible. To fix our ideas we shall assume throughout that J has to be *maximised*: this is of course trivial, since minimising J is the same as maximising $(-J)$, but is in accord with the generic title 'hill-climbing', where J represents height above some datum level, the 'parameters' being horizontal displacement in two specified directions: the mountaineer is in thick fog, has no contour maps, but is provided with an altimeter and a compass. He is required to reach a mountain top as quickly as possible (there may of course be several, but, unless he explores over a wide range he will generally aim at the nearest one). To atone for his various handicaps, he may be allowed, by some miraculous agency, to transport himself to some relatively remote point (specified in terms of the two horizontal co-ordinates) and assess the height at this new point. The problem may be complicated by the constraint that he is not allowed outside some region of his co-ordinate field: this corresponds in practice to limits being placed upon the values of the parameters themselves or upon some quantity in the system which is itself a function of these parameters; in such a case there may be no 'mountain top' within the permitted zone, in which case the highest point will invariably be found on its boundary. In some problems there is the further complication that the whole 'landscape' is changing with time: consider for instance the problem of training a parabolic radio aerial onto a moving celestial body (invisible of course!) emitting a radio bleep signal.

In this introductory Chapter we shall not consider these last two complications (a bounded region and a moving landscape) but we shall consider more than two independent parameter-variables, except of course for the purposes of diagrammatic illustration.

14.1.1 Classification of methods

We shall denote the various parameters by the parameter vector $\boldsymbol{p} = (p_1, p_2, \ldots, p_n)'$. The scalar criterion function J is some nonlinear function of these parameters: $J = f(p_1, p_2, \ldots, p_n) = f(\boldsymbol{p})$. *All* methods suggested (of which we can only give a sample) suppose some starting point $\boldsymbol{p} = \boldsymbol{p}_0$ at which $J = f(\boldsymbol{p}_0) = J_0$ is assessed. Thereafter the majority of useful methods seek, by certain rules, to generate a sequence of values of $\boldsymbol{p}$, say $\boldsymbol{p}_1, \boldsymbol{p}_2, \ldots$ which represent an improvement in the value of J, i.e. $f(\boldsymbol{p}_{i+1}) \geqslant f(\boldsymbol{p}_i)$. It is useful in this connection to visualise the transition from $\boldsymbol{p}_i$ to $\boldsymbol{p}_{i+1}$ as a relation of the form

$$\boldsymbol{p}_{i+1} = \boldsymbol{p}_i + h_i \boldsymbol{d}_i \tag{14.1}$$

in which h_i represents a step-length and $\boldsymbol{d}_i$ is a unit direction-vector: if $\boldsymbol{d}_i$ is well-chosen, h_i will be positive, and it is tempting then to increase the step-length.

We distinguish between *direct search methods*, which rely entirely on the values of J assessed at various points, but making use of any information obtained by assessing J at earlier points, and *gradient methods*, which select a desirable value of $\boldsymbol{d}_i$ in eqn. 14.1 by calculating the values of the partial derivatives of f with respect to each parameter at the point $\boldsymbol{p}_i$. Note however that this distinction is not very sharp, since direct search methods, used to investigate J at points neighbouring $\boldsymbol{p}_i$, may be used to obtain approximate values of the partial derivatives. Note also that true gradient methods can only be used if a mathematical model of the system is available so that the function f is a known mathematical function (assumed differentiable with respect to its arguments).

We also distinguish between the simple case $n = 1$, the so-called *univariate* case, and the case $n > 1$, or *multivariate* case. In the former, since there is only one parameter, the question of gradients barely arises since we either increase this parameter or we decrease it.

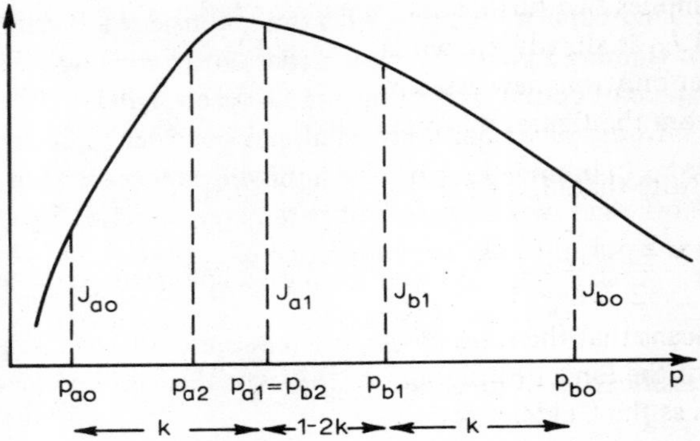

Fig. 14.1 'Golden Section' search

14.2 Direct search methods

14.2.1 Univariate search methods

Although the problem is here greatly simplified by the fact that f is a function of a single variable, the methods used find useful extensions in the field of multivariate search.

(*a*) *'Golden section' and 'Fibonacci' search* We have $J = f(p)$ where p is a single scalar parameter. We suppose that by some common sense physical argument it is known that J has one and only one maximum (and no other turning value) in some range $p_{a0} < p < p_{b0}$. We wish gradually to narrow this range until the maximum is 'boxed in' with sufficient accuracy.

The first step is of course to assess J at the end points of the range, thus obtaining $J_{a0} = J(p_{a0})$ and $J_{b0} = J(p_{b0})$. Clearly the assessment of J at *one* intermediate point will not help to restrict the range within which the maximum is known to lie: both methods require the assessment of J at *two* intermediate points, $p = p_{a1}$ and $p = p_{b1}$, the restriction in range being symmetric in the sense that $p_{a1} - p_{a0} = p_{b0} - p_{b1} = k(p_{b0} - p_{a0})$ say, where $k < \frac{1}{2}$ and where we may imagine the original range, $(p_{b0} - p_{a0})$, to be unit length, to some scale. Then the function J is evaluated at these new points: clearly if $J_{a1} > J_{b1}$, the maximum must lie between J_{a0} and J_{b1}, but if $J_{a1} < J_{b1}$, it must lie between J_{a1} and J_{b0}. For purposes of demonstration we assume the former. (Fig. 14.1).

We may now repeat the process and similarly find two new points p_{a2} and p_{b2} in the range p_{a0} to p_{b1}, using the same value of k as before;

but this implies *two* further assessments of J. Since p_{a1} is already in this range and J_{a1} is already known, it is desirable to make p_{a1} coincide with p_{b2} so that only *one* new assessment of J (at p_{a2}) is necessary. As may be seen from the figure, this requires $k = (1-2k)/(1-k)$, i.e. $k = \frac{3-\sqrt{5}}{2}$. Note that $(1-k) = \frac{\sqrt{5}-1}{2}$ and further that $\frac{k}{1-k} = \frac{3-\sqrt{5}}{\sqrt{5}-1} = \frac{\sqrt{5}-1}{2} = \frac{1-k}{1}$ so that dividing a line in the ratio of k to $(1-k)$ means that the ratio of the shorter segment to the longer equals the ratio of the longer to the sum of the two: this was known to Greek geometers as the Golden Section.

Using this method means that at every stage of range reduction except the first, J need only be assessed at one new point. Moreover the range is reduced in the ratio $(1-k) \doteq 0{\cdot}618$ at every stage. Note however that if at any stage it is found that the values of J at the section points of a range (e.g. p_{a1} and p_{b1}) are equal, then the maximum must lie within this smaller range; if the values are *nearly* equal it is highly probable that this is still the case: in this situation this narrower range may be used to advantage as a new starting range, even though, at the next range reduction *two* new values of J must be obtained.

The Golden Section method uses the same value of k throughout. Suppose that at some stage in the reduction process we use a value k_i and at the next stage a value k_{i+1}. Then to secure the same reduction in the number of J-assessments as previously now requires $k_{i+1} = (1-2k_i)/(1-k_i) = 1 - k_i/(1-k_i)$ which may be written $(1-k_{i+1}) = \frac{1}{1-k_i} - 1$.

Suppose now that successive values of $(1-k_i)$, the range reduction ratio at stage (i), are the ratios of successive numbers in a monotonically decreasing series of positive numbers, . . . , n_{i-1}, n_i, n_{i+1}, . . . , so that $1-k_i = n_i/n_{i-1}$, $1-k_{i+1} = n_{i+1}/n_i$. We deduce $n_{i+1}/n_i = n_{i-1}/n_i - 1$ or $n_{i-1} = n_i + n_{i+1}$. In other words the decreasing series is such that any number in it is the sum of the next two, or, if we write the series backwards, any number is the sum of the previous two. Now the latter is precisely the law of formation of a well-known series attributed to the Italian mathematician Leonardo Fibonnaci*: this series, denoted by the numbers F_i, is formed by $F_i = F_{i-1} + F_{i-2}, i > 2$, starting from $F_1 = F_2 = 1$ to form the series 1, 1, 2, 3, 5, 8, 13, 21, . . . Since however in

* Otherwise known as Leonardo da Pisa; flourished at beginning of 13th century and was largely responsible for the introduction of Arabic numerals into Western Europe – for which we should be grateful!

our range-reduction process we are running through the series backwards, we must, in order to know what value of reduction ratio to start with, know how many stages of reduction we are contemplating: for instance if we start with $1 - k_1 = 13/21$, then $1 - k_2 = 8/13$, $1 - k_3 = 5/8$, $1 - k_4 = 3/5$, $1 - k_5 = 2/3$, and this is clearly the last useful value to use, for making $1 - k_6 = 1/2$ does *not* reduce the range in this ratio since the two points of section coincide and the value of J at this centre point gives no further information on range. With five stages of reduction we therefore start with $1 - k_1 = F_7/F_8$ and, more generally, with N projected stages of reduction we start with $1 - k_1 = F_{N+2}/F_{N+3}$.

A comparison of the two methods shows that after N stages of reduction the Golden Section method reduces the range by the factor $(1 - k)^N = (0{\cdot}61803\ldots)^N$ whereas the Fibonacci method gives a factor $\dfrac{2}{F_{N+3}}$. It will be found that the advantage is always slightly in favour of the Golden Section method; however, if the Fibonacci method *is* taken one stage further than indicated above, with a $(1 - k_{N+1}) = \frac{1}{2}$, and if then one additional computation of J is made *very near* the centre point of the range, the range will be *approximately* halved, giving a factor $\dfrac{1}{F_{N+3}}$ for $(N + 1)$ stages as compared with $(1 - k)^{N+1}$ and this time the advantage is, rather more noticeably, with the Fibonacci method (about 0·85 of the Golden Section method for $N = 5$ or more). The Fibonacci method has the disadvantage that N has to be fixed in advance, though since the reduction ratio on the range is known to be $2/F_{N+3}$ (or $1/F_{N+2}$) for N stages this is not a great handicap.

(*b*) *Method of quadratic approximation to J*. These methods make use of the fact that near a maximum (or minimum) any 'smooth' function of p may be approximated to closely by a parabola, i.e. a quadratic function of p. This parabola, in turn, is completely specified by any three points in the neighbourhood of, and preferably straddling, the maximum. A method suggested by Davies, Swann and Campey[1] is as follows.

Start with some $p = p_0$ and assess J_0. Change p by some step-length h to give $p_1 = p_0 + h$ and assess J_1. Assume for the present that $J_1 > J_0$. Double the step-length to give $p_2 = p_1 + 2h$ and compute J_2; if $J_2 > J_1$, double the step-length again and repeat until some point p_{i+1} is reached such that $J_{i+1} < J_i$. The maximum has now been overshot. Take p_{i+2} midway between p_i and p_{i+1} and assess J_{i+2}. The points $p_{i-1}, p_i, p_{i+2}, p_{i+1}$ constitute four equispaced points straddling the maximum; as in (*a*) above, the maximum lies in the range p_{i-1} to p_{i+2} or p_i to p_{i+1}, depending on whether $J_i \gtrless J_{i+2}$, so that either the point p_{i+1} or the point p_{i-1} may be discarded and the remaining three points used to determine a parabolic approximation to J in the neighbourhood of its maximum.

If, at the first step, $J_1 < J_0$, take the step h in the opposite direction to form, say, $p_{-1} = p_0 - h$. If $J_{-1} > J_0$, double the step-size etc. as before. If by any chance both $J_1 < J_0$ *and* $J_{-1} < J_0$, then the maximum lies between p_1 and p_{-1} and J_1, J_0 and J_{-1} are all assessed.

Thus, whatever the circumstances, the maximum lies in the range of three equispaced points. Denoting these for brevity by $p_a = p_b - H$, p_b and $p_c = p_b + H$ and the corresponding values of J by J_a, J_b and J_c, and assuming moreover that $J = \mathrm{A}p^2 + \mathrm{B}p + \mathrm{C}$ in this neighbourhood, we obtain, on substituting for p and J, three equations for A, B and C: the maximum value of J occurs at $p_m = -\mathrm{B}/2\mathrm{A}$ and $J_m = \mathrm{C} - \mathrm{B}^2/4\mathrm{A}$, A being negative since the parabola is convex upwards. Explicitly, one finds that

$$p_m = p_b + \frac{H}{2}\frac{J_c - J_a}{2J_b - J_a - J_c} \quad \text{and} \quad J_m = J_b + \frac{(J_c - J_a)^2}{8(2J_b - J_a - J_c)} \tag{14.2}$$

14.2.2 Multivariate search methods

(a) The simplex method

This method, originally put forward by Spendley *et al.* in 1962 was presented in an improved form by Nelder and Mead[2] in 1965. The method is briefly as follows.

A simplex, first of all, is an assembly of $(n + 1)$ points, say $\boldsymbol{p}_1, \boldsymbol{p}_2, \ldots, \boldsymbol{p}_n, \boldsymbol{p}_{n+1}$, in the n-dimensional parameter-space, such that det $\begin{bmatrix} p_1 & p_2 & \cdots & p_{n+1} \\ 1 & 1 & \cdots & 1 \end{bmatrix}$ does not vanish: this condition ensures that three points in a plane are not collinear, that four points in space are not coplanar etc. The initial set of points selected may, but need not, be equispaced (equilateral triangle, regular tetrahedron etc.). We suppose that the criterion function is $J = f(\boldsymbol{p})$ and is to be maximised. The rules for modifying the simplex stage by stage with a view to making it converge towards the point $\bar{\boldsymbol{p}}$, say, at which J is maximised will now be given.

J is first assessed at each point of the simplex and we suppose that $\boldsymbol{p}_s$, $\boldsymbol{p}_n$ and $\boldsymbol{p}_b$ are those points in the simplex for which J is *s*mallest, *n*ext smallest and *b*iggest respectively; moreover let $\boldsymbol{p}_g$ be the centre of gravity of the simplex excluding $\boldsymbol{p}_s$. The first change is to reflect $\boldsymbol{p}_s$ in $\boldsymbol{p}_g$ using a reflection coefficient α; the reflected point $\boldsymbol{p}_r$ is given by

$$\boldsymbol{p}_r = \boldsymbol{p}_g + \alpha(\boldsymbol{p}_g - \boldsymbol{p}_s) \quad \text{with} \quad \alpha > 0.$$

Assess J_r. If J_r lies between J_b and J_n, the point $\boldsymbol{p}_r$ replaces $\boldsymbol{p}_s$ to form a new simplex and the process is repeated, i.e. we again reflect the point with smallest J-value in the centroid of the remaining points. If, however, $J_r > J_b$ so that a new maximum has been created at $\boldsymbol{p}_r$, we argue that this direction of travel is a good one and the distance travelled is increased by an *e*xpansion coefficient γ to give

$$\boldsymbol{p}_e = \boldsymbol{p}_g + \gamma(\boldsymbol{p}_r - \boldsymbol{p}_g) \quad \text{with} \quad \gamma > 1,$$

creating a new point on the line $\boldsymbol{p}_s\boldsymbol{p}_g\boldsymbol{p}_r$ extended beyond $\boldsymbol{p}_r$. If $J_e > J_b$ the expansion is considered a success and $\boldsymbol{p}_e$ replaces $\boldsymbol{p}_s$ in the next simplex. If not, the expansion is a failure and $\boldsymbol{p}_r$ replaces $\boldsymbol{p}_s$.

Finally, if $J_r < J_n$ the reflected point $\boldsymbol{p}_r$ is the minimal point for the proposed new simplex and would, following the normal rule, be re-reflected in $\boldsymbol{p}_g$, probably an unfruitful operation. Instead, this situation is dealt with by a *c*ontraction coefficient β, used in one of two ways: if $J_r > J_s$, $(\boldsymbol{p}_r - \boldsymbol{p}_g)$ contracts and the result of the contraction is given by

$$\boldsymbol{p}_c = \boldsymbol{p}_g + \beta(\boldsymbol{p}_r - \boldsymbol{p}_g) \quad \text{with} \quad 0 < \beta < 1,$$

but if $J_r < J_s$ then $\boldsymbol{p}_s$ replaces $\boldsymbol{p}_r$ to give

$$\boldsymbol{p}_c = \boldsymbol{p}_g + \beta(\boldsymbol{p}_s - \boldsymbol{p}_g).$$

Provided that $J_c > J_s$, the contraction process is considered a success and $\boldsymbol{p}_s$ is replaced by $\boldsymbol{p}_c$ in the new simplex; if however $J_c < J_s$, the contraction is a failure and in this event a new simplex is created by retaining $\bar{\boldsymbol{p}}_b$ only and halving the distance from $\boldsymbol{p}_b$ of every other point in the simplex.

Suggested values of α, β and γ are $\alpha = 1$, $\beta = \frac{1}{2}$, $\gamma = 2$, but these values may be modified by the investigator in the light of his experience. Note that if expansion is used, the reflection coefficient α is effectively modified to $\alpha\gamma$; similarly if contraction is used, α is effectively modified to $+\alpha\beta$ or $-\alpha\beta$ according as $J_r \gtrless J_s$. These modifications of the reflection coefficient, however, should only be used according to the rules given above.

The method is probably one of the most successful direct search techniques suggested. Its use in approaching a 'mountain top' is shown in Fig. 14.2, necessarily limited to two variable parameters. Only the first few stages of the search are shown, starting from the simplex A, B, C. Note that the first two stages in modifying the simplex result in

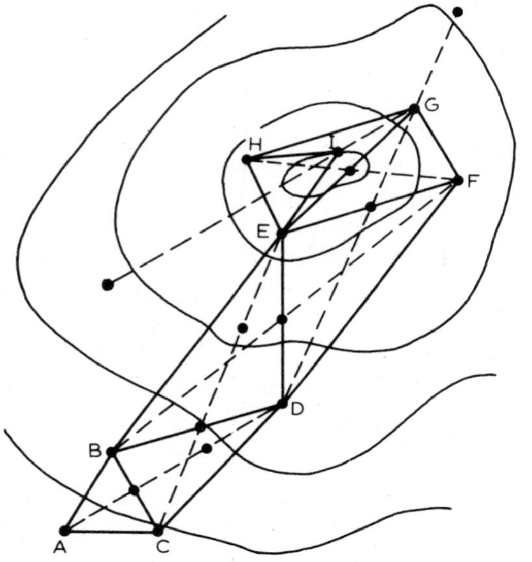

Fig. 14.2 The simplex method in a plane ($n = 2$). D and E are obtained by the expansion rule, G and I by the contraction rule, first and second versions, respectively. The process may of course be continued beyond the simplex E, H, I, shown.

successful expansions, the third stage is a normal reflection, the fourth stage calls for a contraction of the first kind, the fifth is again a normal reflection, the sixth is a contraction of the second kind.

(*b*) *Cyclical variable search methods*

Basically this seems the most obvious method of attack: we explore the value of J changing only one variable at a time until J is maximised for that particular variable; for this purpose any univariate search method may be used. This search is continued for each of the n variables and this cycle of n univariate searches is then repeated until the changes in J are considered negligible. Since each univariate search itself requires a number of J assessments, the process is liable to be lengthy. It is illustrated in Fig. 14.3 (full-line trace) again necessarily for only two variables. Each univariate search (with respect to p_1 or p_2) terminates at a point where the line of search is tangential to a contour.

It will be noted that if, with the same starting point, we had taken search directions roughly parallel to the 'principal axes' of the approximately elliptical contours (a search path shown in broken line in Fig. 14.3) the steps would have been appreciably larger and their number correspondingly smaller. But of course the searcher does not know the

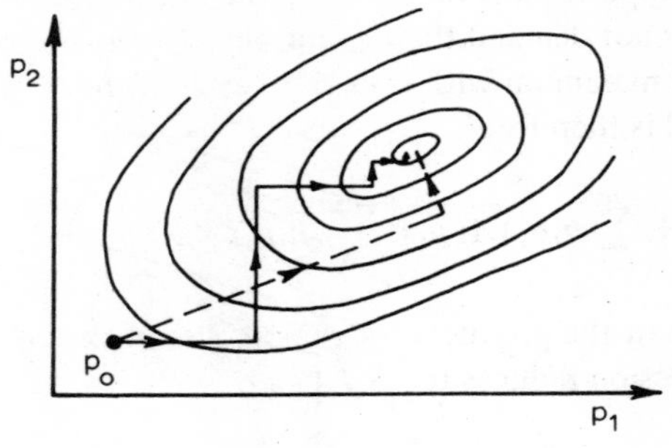

Fig. 14.3 Alternating-variable search. The full-line trace corresponds to varying p_1 and p_2 alternately, the broken-line trace corresponds to directions roughly parallel to the axes of symmetry of the contours

shape of the contours, which, in any case, may not be of the elliptical form shown. Nevertheless, a number of methods have been suggested which seek to abbreviate the search process by changing the directions of search from directions parallel to the co-ordinate axes to more rewarding directions as the search proceeds. For a fuller discussion of such methods the reader is referred to Reference 3. We shall only, in this context, draw attention to the properties of *conjugate directions*.

If J is a maximum at a discrete *point* in the parameter-space (i.e. does not attain its maximal value on a *line* of points, or a *plane* etc.) then, by analogy with the univariate case, in the neighbourhood of the maximal point, J may be approximated to by a quadratic expression in $\boldsymbol{p}$:

$$J = \boldsymbol{p}'\boldsymbol{A}\boldsymbol{p} + \boldsymbol{b}'\boldsymbol{p} + c$$

where $\boldsymbol{A}$ is nonsingular and may, as in all quadratic forms, be assumed symmetric, since any skew-symmetric component of $\boldsymbol{A}$ contributes zero to $\boldsymbol{p}'\boldsymbol{A}\boldsymbol{p}$. Two direction vectors, $\boldsymbol{q}_i$ and $\boldsymbol{q}_j$, assumed here of unit norm, are defined as conjugate with respect to $\boldsymbol{A}$ if $\boldsymbol{q}_i'\boldsymbol{A}\boldsymbol{q}_j = 0$. (Since $\boldsymbol{A}$ is symmetric and $\boldsymbol{q}_i'\boldsymbol{A}\boldsymbol{q}_j$ is a scalar and therefore equal to its transpose, it follows that $\boldsymbol{q}_j'\boldsymbol{A}\boldsymbol{q}_i = 0$ also, so that conjugacy is a mutual property). Suppose then that $\boldsymbol{q}_1, \boldsymbol{q}_2, \ldots, \boldsymbol{q}_n$ form a linearly independent set of direction-vectors, conjugate with respect to $\boldsymbol{A}$ and therefore such that for any $i \neq j$ in the range $1 \leqslant (i,j) \leqslant n$, $\boldsymbol{q}_i'\boldsymbol{A}\boldsymbol{q}_j = 0$. With any starting point $\boldsymbol{p}_0$ in the parameter space, we may suppose that the maximal point $\bar{\boldsymbol{p}}$ (where J is a maximum) may be reached in n steps by suitably chosen displacements h_i in the n directions $\boldsymbol{q}_i$ $(i = 1, 2, \ldots, n)$ i.e.

$$\bar{\boldsymbol{p}} = \boldsymbol{p}_0 + \sum_{i=1}^{n} h_i \boldsymbol{q}_i$$

Note that this supposition is validated by the linear independence of the $\boldsymbol{q}_i$ only and does not demand their conjugacy. The h_i are to be chosen so that $J(\bar{\boldsymbol{p}})$ is a maximum with respect to each of the h_i. The function to be maximised is therefore

$$J(\bar{\boldsymbol{p}}) = (\boldsymbol{p}_0' + \sum_{i=1}^{n} h_i \boldsymbol{q}_i')\boldsymbol{A}(\boldsymbol{p}_0 + \sum_{i=1}^{n} h_i \boldsymbol{q}_i) + \boldsymbol{b}'(\boldsymbol{p}_0 + \sum_{i=1}^{n} h_i \boldsymbol{q}_i) + c$$

On multiplying out the products and making use of the conjugacy property, this expression reduces to

$$J(\bar{\boldsymbol{p}}) = J(\boldsymbol{p}_0) + \sum_{i=0}^{n} (h_i \boldsymbol{q}_i' \boldsymbol{A} \boldsymbol{p}_0 + \boldsymbol{p}_0' \boldsymbol{A} h_i \boldsymbol{q}_i + h_i^2 \boldsymbol{q}_i' \boldsymbol{A} \boldsymbol{q}_i + \boldsymbol{b}' h_i \boldsymbol{q}_i)$$

$$= J(\boldsymbol{p}_0) + \sum_{i=0}^{n} \{h_i^2 . \boldsymbol{q}_i' \boldsymbol{A} \boldsymbol{q}_i + h_i \boldsymbol{q}_i'(2\boldsymbol{A}\boldsymbol{p}_0 + \boldsymbol{b})\}$$

$$= J(\boldsymbol{p}_0) + \sum_{i=0}^{n} g_i(h_i), \text{ say.}$$

This function is clearly maximised with respect to the h_i when every term $g_i(h_i)$, $i = 1, 2, \ldots, n$, is maximised with respect to the *single* h_i of which it is a quadratic function. In other words, the required h_i may be found by a *single* univariate search along each of the directions $\boldsymbol{q}_i$ in turn, using the maximal point of any such search as the starting point for the next: the maximal point $\bar{\boldsymbol{p}}$ is therefore reachable in n steps and no repetition of the step-cycle is necessary.

We do not of course, in practice, know the value of $\boldsymbol{A}$ or $\boldsymbol{b}$, nor the values of the various $\boldsymbol{q}_i$. A number of search methods, however, use a preliminary search to determine n conjugate directions, based merely on the assumption that J is quadratic in $\boldsymbol{p}$ (i.e. that the search is being conducted in the neighbourhood of $\bar{\boldsymbol{p}}$) and without knowing $\boldsymbol{A}$, $\boldsymbol{b}$ and c. The best of these methods appears to be due to Powell.[4]

14.3 Gradient methods

14.3.1

If at any point $\boldsymbol{p}$ in the parameter space we are able to determine, exactly or approximately, the partial derivatives of J with respect to each parameter in turn, i.e. if we can find $\dfrac{\partial J}{\partial p_i}$ $(i = 1, 2, \ldots, n)$, then the vector with components $\frac{\partial J}{\partial p_1}, \frac{\partial J}{\partial p_2}, \ldots, \frac{\partial J}{\partial p_n}$ acts in such a direction that *for a given, small displacement, J increases more in this direction than in any other*. This vector is in fact the gradient of the scalar J with respect to the vector $\boldsymbol{p}$. The above property is easily proved. For let p_i increase by $\delta p_i (i = 1, 2, \ldots, n)$; then J increases by

$$\delta J = \sum_{i=1}^{n} \frac{\partial J}{\partial p_i} . \delta p_i$$

Also, since the displacement δs is of given length, we have

$$(\delta s)^2 = \sum_{i=1}^{n} (\delta p_i)^2 = \text{constant}$$

Since δJ is to have a maximum while $(\delta s)^2$ is to be constant, we deduce

$$\sum_{i=1}^{n} \frac{\partial J}{\partial p_i} . d(\delta p_i) = 0 \quad \text{and} \quad \sum_{i=1}^{n} \delta p_i . d(\delta p_i) = 0$$

equations which must be satisfied for arbitrary values of the several increments $d(\delta p_i)$. It follows at once that

$$\frac{\delta p_1}{\left(\frac{\partial J}{\partial p_1}\right)} = \frac{\delta p_2}{\left(\frac{\partial J}{\partial p_2}\right)} = \ldots = \frac{\delta p_n}{\left(\frac{\partial J}{\partial p_n}\right)} = k \text{ say, so that } \delta J = k \sum_{i=1}^{n} \left(\frac{\partial J}{\partial p_i}\right)^2$$

and k must therefore be positive since δJ is positive.

The direction of the resultant gradient is therefore clearly a good direction to move in. Unfortunately, of course, the gradient varies as the travel proceeds (tending to zero at the maximal point) so that we have the option of either taking very small steps and re-assessing the gradient at every step (an accurate but very laborious method of reaching the maximal point) or taking larger steps, at the expense of possibly following a more devious path to the optimal point, but with the advantage of fewer J assessments.

A straightforward if rather crude method of using gradient methods is attributed to Cauchy: assess the gradient at the starting point, $\boldsymbol{p}_0$, and travel in this direction an arbitrary step-length to reach $\boldsymbol{p}_1 = \boldsymbol{p}_0 + h\boldsymbol{g}_0$ where h is the length of the step and $\boldsymbol{g}_0$ is the gradient at $\boldsymbol{p}_0$. Provided that $J_1 > J_0$, assess $\boldsymbol{g}_1$ and repeat to find $\boldsymbol{p}_2 = \boldsymbol{p}_1 + h\boldsymbol{g}_1$. If however $J_1 < J_0$ then the selected step-length h is too large and is halved, or quartered etc. until $J_1 > J_0$. The same rules are followed at every stage.

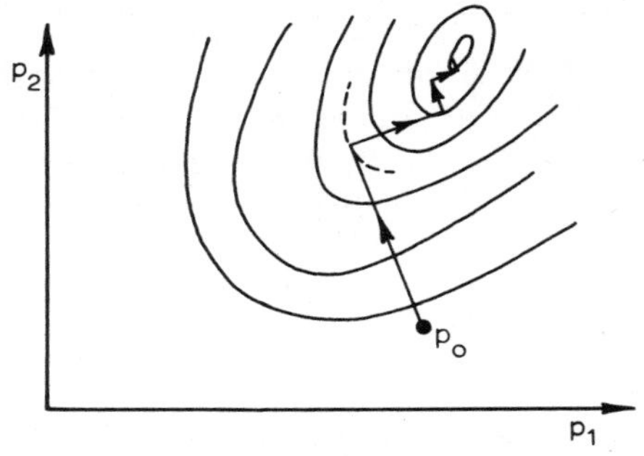

Fig. 14.4 Method of steepest ascent, combined with univariate searches for a maximum at each stage

A more sophisticated method, generally called the *method of steepest ascent*, is, starting from $\boldsymbol{p}_0$, to conduct a univariate search in the direction $\boldsymbol{g}_0$ until a maximum is found; at this point the gradient is re-assessed, a further univariate search for a maximum is carried out etc., etc. The method is illustrated in Fig. 14.4, again, necessarily, for only two variable parameters. It is to be noted that the gradient, at any point in this diagram, has a direction along the normal to the contour through that point, whereas at the end of a step, the direction of travel is a tangent to the contour through the end-point. The two directions used are therefore two fixed, mutually perpendicular directions and, in general, the method does not appear to be a great advance on the method of cyclical variables illustrated in Fig. 14.3.

Note that the constancy of the two directions of travel is limited to the case $n = 2$: if $n > 2$, say $n = 3$, although any two *successive* directions of travel will still be perpendicular (one in the tangential plane to a contour surface, the other along the normal to that surface), the third direction, though perpendicular to the second, is not in general perpendicular to the first, nor, therefore, is the fourth in general parallel to the first, so that the cycle of directions does not in general repeat.

Other methods based on gradient assessment are to be found in the literature[3] but do not appear to produce major improvements and are not discussed here. We should point out, however, that provided the system is represented by a mathematical model so that J is a *known* function of the variable parameters $p_1, p_2, \ldots, p_n$, then the assessment of the gradient at any point in the parameter-space merely involves the computation of the n partial derivatives $\dfrac{\partial J}{\partial p_i}$ $(i = 1, 2, \ldots, n)$, all of which are obtainable as mathematical functions. But in the absence of a reliable mathematical model, the gradient can only be assessed, approximately, by varying each parameter in turn by a small but finite amount and, after steady-state conditions are established, measuring the resulting change in J; since this change in J will also be small and since its measurement is liable to be vitiated by noise and other disturbances, accuracy is highly unlikely. In this situation therefore, gradient methods are not recommended. There is, however, one aspect of experimental gradient assessment which deserves mention, namely perturbation techniques.

14.4 Perturbation theory

The essence of the method lies in the following piece of elementary analysis. Suppose at first, for simplicity, that we are dealing with the univariate problem, so that J is some unknown function of a single variable p. Let the system be functioning under steady-state conditions with $p = p_0$. Superimpose on this value of p a sinusoidal perturbation ($a \sin kt$), of *small* amplitude a. Then it is arguable that $J(p_0)$ is thereby modified to a perturbed value

$$J_P = J(p_0) + \frac{\partial J}{\partial p} . a \sin kt + \frac{\partial^2 J}{\partial p^2} . \frac{a^2 \sin^2 kt}{2!} + \ldots$$

the derivatives being evaluated at $p = p_0$. If we now form the product of J_P and a sinusoid co-phasal with the perturbation, say $b \sin kt$, we have $bJ_P \sin kt = bJ(p_0) \sin kt + \frac{\partial J}{\partial p} .ab \sin^2 kt + \frac{\partial^2 J}{\partial p^2} . \frac{a^2 b \sin^3 kt}{2!} + \ldots$ and if we now take the average value of this product over one or more complete cycles of the perturbation, we deduce

$$q_0 \equiv (bJ_P \sin kt)_{mean} = \frac{\partial J}{\partial p} . \frac{ab}{2} + \ldots \tag{14.3}$$

succeeding terms being of order $a^3, a^5, \ldots$ and therefore in general negligible provided that a is reasonably small. The output q_0 of the process is therefore a measure of the gradient of J with respect to p at the operating point $p = p_0$. If q_0 is therefore added to p_0, the system will operate with $p = p_1 = p_0 + q_0$ and p_1 will be closer to the value of p, p_{opt}, which maximises J, than to p_0. With the same perturbation as before we now obtain $q_1 = \frac{ab}{2} \left(\frac{\partial J}{\partial p}\right)_{p=p_1}$, which is added to p_1 to give $p_2 = p_1 + q_1$ etc. and the process is continued until p approaches p_{opt} sufficiently closely.

If we suppose for instance that J is quadratic in p, we may write

$$J = J_{max} - c(p - p_{opt})^2 \quad \text{and} \quad \frac{\partial J}{\partial p} = 2c(p_{opt} - p) \qquad (c > 0)$$

Hence, for $r = 0, 1, 2, \ldots$, we deduce

$$q_r = abc(p_{opt} - p_r) \quad \text{and} \quad p_{r+1} = p_r + q_r$$

Hence $p_{r+1} = p_r + abc(p_{opt} - p_r)$, or, subtracting each side from p_{opt}:

$$p_{opt} - p_{r+1} = (1 - abc)(p_{opt} - p_r)$$

showing that the successive deviations from $p = p_{opt}$ form a geometrical progression and therefore tend to zero provided only that $abc < 2$, i.e. b should not be made too large. (Note that if we could make $abc = 1$, then p_1 would equal p_{opt} and the system would operate optimally after one step; but since we do not know the value of c, we are unable to adjust ab to the value $1/c$).

The principle may at once be extended to the multivariate case: each parameter is subjected to a sinusoidal perturbation, one at a time, and successive improvements made to each parameter in turn, the cycle being repeated until the value of J is approximately stationary.

If it is possible, physically, to make every signal q_r modify p_r to p_{r+1} automatically (through some actuator) then we shall have achieved a step-by-step *adaptive* system which will *automatically* adjust its parameter values until the criterion function J is maximised. We stress the *step-by-step* nature of the adjustments. We shall *not* achieve the desired result by a continuous feed-back from q to p: for suppose we do this, starting as before with, in the univariate problem, $p = p_0$. Then the feed-back makes $p = p_0 + q$, say, where $q = \frac{ab}{2}\left(\frac{\delta J}{\delta p}\right)_{p = p_0 + q}$, from eqn. 14.3. The system will operate at whatever value of q, satisfies this equation, but this *cannot be* optimally, for if $p_0 + q = p_{opt}$, the derivative term must vanish so that q must vanish and then $p = p_0 \neq p_{opt}$.

14.4.1

In conclusion we may note a number of practical difficulties. In the first place, J, although an unknown function of the parameters, must be measurable and capable of expression, through some transducer, as an electrical signal, which can then be subjected to phase-sensitive rectification and smoothing to produce the q signal. Secondly, it may not be easy to subject a parameter to sinusoidal fluctuations; for example if the parameter in question is a steam-valve! Thirdly, the theory assumes that J is a *static* function of the parameters: actually J may, and usually is, related to the parameters by *dynamic* relations, so that even the steady-state value of J_P is a function not only of p_0, a and $\sin kt$ but also of k explicitly. This is not serious since it merely modifies the coefficient of $\frac{\partial J}{\partial p}$ in eqn. 14.3 without affecting the form of the relation.

14.5 References

1 SWANN, W.H.: 'Report on the development of a new direct search method of optimisation', I.C.I. Ltd. Central Instrument Laboratory Research Note 64/3, 1964

2 NELDER, J.A. and MEAD, R.: 'A simplex method for function minimization', *Comput. J.*, 1964, 7, pp. 308–13

3 I.C.I. Monograph No. 5: Non-linear Optimization Techniques.

4 POWELL, M.J.D.: 'An efficient method of finding the minimum of a function of several variables without calculating derivatives', *Comput. J.*, 1964, 7, pp. 155–62

14.6 Examples 14

Note: Although in these examples the function to be optimised, J, is given as an explicit function of relevant variables, in practice the values of J have to be found experimentally, its functional form being supposed unknown to the investigator; in these theoretical examples experimental determination is necessarily replaced by computation.

1. If $J(x)$ is a function of the single variable x given by

$$J(x) = \log_e(1+x)/(1+2x)^3$$

find the value of x in the range $0 < x < 1$ which maximises J, using (*a*) the method of the golden section (*b*) the Fibanocci method, locating this value of x within a range of 0·02. Verify the value obtained by normal differentiation procedure.

2. If $J(x_1, x_2) = (1-x_1)^2 + 10(x_2 - x_1^2)^2$, use the Nalder and Mead simplex method to locate the values of x_1 and x_2 which minimise J (i.e. maximise its negative), starting from the simplex formed by the

points $(0, 0)$, $(\frac{1}{2}, 0)$, $(\frac{1}{2}, \frac{1}{2})$ in the (x_1, x_2) plane. The correct values should be obvious by inspection.

3. In a simple system, $J(x)$, the function to be maximised, may be approximated to by the quadratic $J(x) = 0{\cdot}15 + x - x^2$, where x, normally a function of time t, satisfies the equation $\dot{x}(t) + x(t) = u(t)$, $u(t)$ being an adjustable parameter.

For a long time prior to $t = 0$, $u = U =$ constant. Find the value of J at $t = 0$, $= J_U$ say.

For $t \geqslant 0$, u is increased by a small sinusoidal perturbation so that, for $t \geqslant 0$, $u = U + \Delta \sin \omega t$. Calculate the resulting values of $x(t)$, and hence of J, for any $t \geqslant 0$.

This value of J is multiplied by a signal cophasal with the perturbation, say $k \sin \omega t$, and the product is filtered by a low-pass filter to eliminate any frequencies of the order of ω or greater. Show that, as $t \to \infty$, this 'smoothed' product approximates to $\dfrac{k\Delta}{1 + \omega^2} \cdot \dfrac{\partial J_U}{\partial U}$ and deduce that if this smoothed product is added to U on a step-by-step basis, the system will ultimately operate with such a value of U that J_U is maximised. What is the effect of the value of k upon this step-by-step approach to the optimum?

Index

(see also 'Table of contents')